Scotland's leading educational publishers

T0173575

National 5
PHYSICS
PRACTICE QUESTION BOOK

N5 PHYSICS
PRACTICE QUESTION BOOK

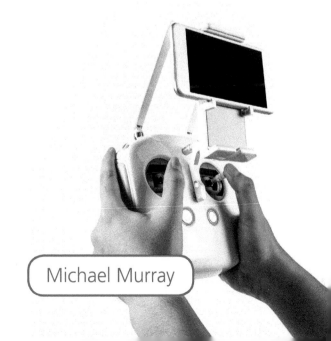

Michael Murray

001/17052018

10 9 8 7 6 5 4 3 2 1

ISBN 9780008263591

Published by
Leckie & Leckie Ltd
An imprint of HarperCollins*Publishers*
Westerhill Road, Bishopbriggs, Glasgow, G64 2QT
T: 0844 576 8126   F: 0844 576 8131
leckieandleckie@harpercollins.co.uk   www.leckieandleckie.co.uk

Commissioning editor: Kerry Ferguson
Project manager: Rachel Allegro

Special thanks to
Jouve (layout)
Ink Tank (cover)
Louise Robb (copyedit)
Nick Forwood (answer check)
Dylan Hamilton (proofread)

Printed in the UK by CPI Group (UK) Ltd, Croydon, CR0 4YY

A CIP Catalogue record for this book is available from the British Library.

Acknowledgements
Whilst every effort has been made to trace the copyright holders, in cases where this has been unsuccessful, or if any have inadvertently been overlooked, the Publishers would gladly receive any information enabling them to rectify any error or omission at the first opportunity.

Leckie & Leckie would like to thank the following copyright holders for permission to reproduce their material:

Cover and page 1: © leolintang / Shutterstock & © Longchalerm Rungruang / Shutterstock

p.1: Sabelskaya / Shutterstock; p.4: SergeV / Shutterstock; p.9: fckncg / Shutterstock; p.15: Inspiring / Shutterstock; p.19 (TL): sirtravelalot / Shutterstock; p.19 (TR): Kamenetskiy Konstantin / Shutterstock; p.19 (BL): Gelpi / Shutterstock; p.19 (BR): Antonio Gravante / Shutterstock; p.22: Natali Snailcat / Shutterstock; p.23 (T): gcafotografia / Shutterstock; p.23 (B): Whitevector / Shutterstock; p.28: M-O Vector / Shutterstock; p.34: T-Kot / Shutterstock; p.48: Aleksei Martynov / Shutterstock; p.50: P.S_2 / Shutterstock; p.53: Clip Art / Shutterstock; p.54: ashva / Shutterstock; p.55: Duda Vasilii / Shutterstock; p.62: Rvector / Shutterstock; p.63 (T): phoelixDE / Shutterstock; p.63 (B): Nerthuz / Shutterstock; p.68: chromatos / Shutterstock; p.69: robuart / Shutterstock; p.70: Sebastian Kaulitzki / Shutterstock

CONTENTS

How to use this book

Welcome to Leckie & Leckie's National 5 Physics Practice Question Book. This book is ideal to use alongside the Leckie & Leckie National 5 Physics Student Book. Questions have been written to provide practice for topics and concepts which have been identified as challenging for many students.

Examples

Examples with worked solutions provide support for particularly tricky concepts.

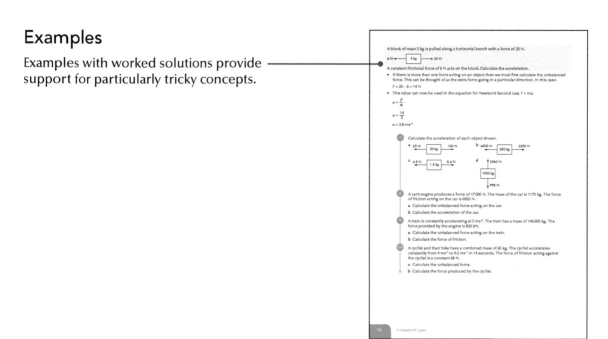

Hints

Where appropriate, hints are provided to help give extra guidance and support.

Answers

Check your own work. The answers are provided online at:

www.leckieandleckie.co.uk/page/Resources

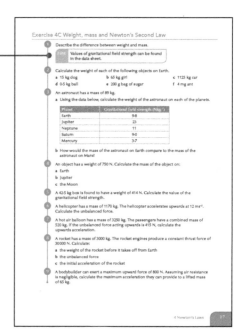

DATA SHEET

Speed of light in materials

Material	Speed in ms^{-1}
Air	3.0×10^8
Carbon dioxide	3.0×10^8
Diamond	1.2×10^8
Glass	2.0×10^8
Glycerol	2.1×10^8
Water	2.3×10^8

Gravitational field strengths

	Gravitational field strength on the surface in N kg^{-1}
Earth	9.8
Jupiter	23
Mars	3.7
Mercury	3.7
Moon	1.6
Neptune	11
Saturn	9.0
Sun	270
Uranus	8.7
Venus	8.9

Specific latent heat of fusion of materials

Material	Specific latent heat of fusion in J kg^{-1}
Alcohol	0.99×10^5
Aluminium	3.95×10^5
Carbon Dioxide	1.80×10^5
Copper	2.05×10^5
Iron	2.67×10^5
Lead	0.25×10^5
Water	3.34×10^5

Specific latent heat of vaporisation of materials

Material	Specific latent heat of vaporisation in J kg^{-1}
Alcohol	11.2×10^5
Carbon Dioxide	3.77×10^5
Glycerol	8.30×10^5
Turpentine	2.90×10^5
Water	22.6×10^5

Speed of sound in materials

Material	Speed in ms^{-1}
Aluminium	5200
Air	340
Bone	4100
Carbon dioxide	270
Glycerol	1900
Muscle	1600
Steel	5200
Tissue	1500
Water	1500

Specific heat capacity of materials

Material	Specific heat capacity in J kg^{-1} °C^{-1}
Alcohol	2350
Aluminium	902
Copper	386
Glass	500
Ice	2100
Iron	480
Lead	128
Oil	2130
Water	4180

Melting and boiling points of materials

Material	Melting point in °C	Boiling point in °C
Alcohol	−98	65
Aluminium	660	2470
Copper	1077	2567
Glycerol	18	290
Lead	328	1737
Iron	1537	2737

Radiation weighting factors

Type of radiation	Radiation weighting factor
alpha	20
beta	1
fast neutrons	10
gamma	1
slow neutrons	3
X-rays	1

RELATIONSHIPS SHEET

$d = vt$

$d = \bar{v}t$

$s = vt$

$s = \bar{v}t$

$a = \dfrac{v - u}{t}$

$F = ma$

$W = mg$

$E_w = Fd$

$E_p = mgh$

$E_k = \dfrac{1}{2}mv^2$

$Q = It$

$V = IR$

$V_2 = \left(\dfrac{R_2}{R_1 + R_2}\right)V_s$

$\dfrac{V_1}{V_2} = \dfrac{R_1}{R_2}$

$R_T = R_1 + R_2 + ...$

$\dfrac{1}{R_T} = \dfrac{1}{R_1} + \dfrac{1}{R_2} + ...$

$P = \dfrac{E}{t}$

$P = IV$

$P = I^2R$

$P = \dfrac{V^2}{R}$

$E_h = cm\Delta T$

$E_h = ml$

$p = \dfrac{F}{A}$

$p_1V_1 = p_2V_2$

$\dfrac{p_1}{T_1} = \dfrac{p_2}{T_2}$

$\dfrac{V_1}{T_1} = \dfrac{V_2}{T_2}$

$\dfrac{pV}{T} = \text{constant}$

$f = \dfrac{N}{t}$

$v = f\lambda$

$T = \dfrac{1}{f}$

$A = \dfrac{N}{t}$

$D = \dfrac{E}{m}$

$H = Dw_r$

$\dot{H} = \dfrac{H}{t}$

1 Vectors and scalars

Exercise 1A Average speed

1. A model train travels at a constant speed of 5 ms⁻¹.

   Calculate the distance it travels in 3 seconds.

2. A car is moving with an average speed of 12·5 ms⁻¹.
   How far will it travel in 120 seconds?

3. A lorry travels along a motorway at a constant speed of 15 ms⁻¹ for 3000 m. How long does it take to cover this distance?

4. In a 50 m freestyle swimming race, a swimmer finishes in a time of 33 seconds. Assuming the swimmer maintains a constant speed throughout the race, calculate the average speed.

5. The Clyde Tunnel is 762 metres long. A car takes 1 minute 10 seconds to travel through the tunnel. Calculate the average speed.

   > **Hint** Remember that 1 km = 1000 m

6. The Bloodhound supersonic car is being constructed in the UK. It aims to break the world land speed record. It will cover a distance of 1 km in 2·1 seconds. Calculate the average speed in ms⁻¹.

7. A transatlantic flight from Glasgow to New York takes 6 hours. The distance travelled by the plane is 5200 km. Calculate the average speed in metres per second.

8. Every year, the Ultra-Trail du Mont Blanc (UTMB) takes place in the Alps. It is considered to be one of the most difficult foot races in the world, covering a distance of 166 km.

   The times taken by the top three finishers are shown below.

Francois	19 hours 2 minutes
Kilian	19 hours 17 minutes
Tim	19 hours 53 minutes

   a Calculate the average speed in metres per second for each runner (give your answers to two decimal places).

   b Calculate the greatest average speed in kilometres per hour.

9 Below is a section of the train timetable from Glasgow to Edinburgh. The length of the journey is 66 km. The average speed of the train is 75 kmh^{-1}.

Departure	Arrival
07:30	a
b	08:38
08:00	c
08:15	d

Calculate the missing arrival and departure times.

10 A cyclist rides along a road.

traffic lights

Describe a method by which the average speed of the cyclist could be measured.

Your description must include the following:

• measurements made

• equipment used

• any necessary calculations

1 A physics trolley is positioned at the top of a slope. A 5 cm card is attached to the trolley. The trolley is released and the card cuts a beam of light at the bottom of the slope. The beam of light is connected to a timer. The timer reading indicates the card took 0·05 seconds to pass through the beam.

Calculate the instantaneous speed of the trolley in metres per second.

2 A trolley is released from rest at the top of a track. A card mounted on the trolley passes through a light gate at the bottom of the track. The following information is recorded:

Length of the card = 40 mm

Distance travelled by the trolley = 1·5 m

Time for the card to pass the light beam = 0·042 s

Calculate the instantaneous speed in metres per second.

3 A pedestrian wants to measure the instantaneous speed of a bus as it passes a set of traffic lights. The pedestrian measures the length of the bus as 12 m. A stopwatch is started when the front end of the bus passes the pedestrian and stopped when the rear of the bus has passed. This time is recorded as 3·8 seconds.

a Calculate the instantaneous speed of the bus.

b Suggest an improvement to the experiment that would make the result more reliable.

4 A car is travelling on the motorway where the speed limit is 50 mph (22 ms⁻¹). When the car passes a speed camera, it takes two photographs that are 0·8 seconds apart. The car travels 15·6 m in this time. Show by calculation if the car is within the speed limit.

5 During a Physics lesson, two students are investigating the motion of a trolley down a slope. They want to measure the instantaneous speed of the trolley as it passes through the light beam.

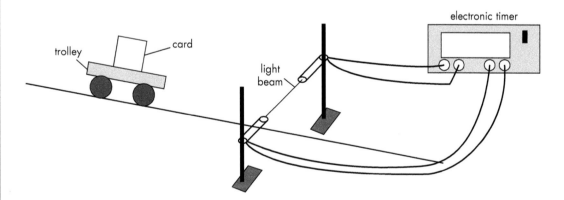

Describe how the students can use the apparatus to measure the instantaneous speed of the trolley.

Your description must include the following:

• measurements made

• equipment used

• any necessary calculations

Describe the difference between vector and scalar quantities.

 Identify the following quantities as either vectors or scalars:

force, speed, velocity, distance, displacement, acceleration, mass, time, energy

> **Hint** Vector quantities must be described by their magnitude **and** direction: 150 m at 37° east of north.

 A student walks 100 metres north before turning and walking 40 metres south. What is the student's resultant displacement?

4 A drone flies 50 metres south, 100 metres west, then 50 metres north. What is the resultant displacement of the drone?

A pupil on their way to school walks 30 m north before turning and walking 40 m west. What is the pupil's resultant displacement?

- Vectors must be joined tip to tail.

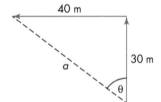

- Use Pythagoras to determine the value of the hypotenuse. This value is the magnitude of the displacement.

$a^2 = b^2 + c^2$

$a^2 = 30^2 + 40^2$

$a^2 = 2500$

$a = 50 \ m$

- As displacement is a vector, it cannot be described by magnitude alone. It must also have a direction. The direction can be calculated using trigonometry (SOH CAH TOA).

$\tan \theta = \dfrac{O}{A}$

$\tan \theta = \dfrac{40}{30}$

$\tan \theta = 1\cdot33$

$\theta = 53°$

- This angle on its own is not specific enough. We need to state the angle either as a bearing or with compass directions. In this case:

$\theta = 53°$ W of N

OR

$\theta = 360° - 53° = 307°$

- Now give your final answer showing the displacement and its direction.

> **Hint** Remember to join vectors tip to tail.

Displacement = 50 m at 53° *W of N* OR 50 m at 307°

5 A long-distance runner travels 300 metres north, then 400 metres east. Using a scale diagram or calculation, determine the resultant displacement.

6 On his way to school, a boy walks 250 metres east, then 400 metres south. Using a scale diagram or calculation, determine the resultant displacement.

7 A bicycle travels 900 metres north, then 750 metres west. The journey takes 420 seconds. Calculate the:

  a total distance travelled **b** resultant displacement

  c average speed **d** average velocity

8 A hillwalker travels 10 km east, then 5 km north. The journey takes 2 hours. Calculate the:

  a total distance travelled by the hillwalker

  b resultant displacement

  c average speed

  d average velocity

9 A boat crosses a river that is flowing at a constant speed of 8 ms^{-1} west. The boat is travelling north at a constant speed of 15 ms^{-1}. Using a scale diagram or calculation, determine the resultant velocity.

10 An aircraft is flying south at a constant speed of 100 kmh^{-1}. The wind is blowing from east to west at a constant speed of 80 kmh^{-1}. Using a scale diagram or calculation, determine the resultant velocity.

2 Velocity–time graphs

Exercise 2A Drawing and interpreting velocity–time graphs

1 Interpret each of the velocity–time graphs below to describe the motion of the object:

a
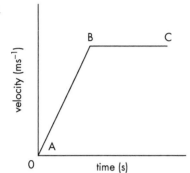

b
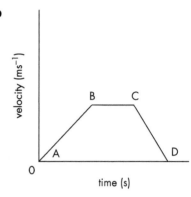

c
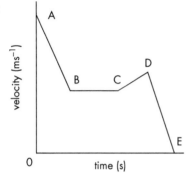

d
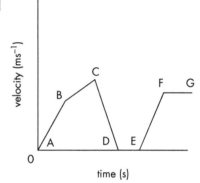

2 Plot a velocity–time graph for each of the data sets below:

a

time (s)	0	5	10	15	20	25	30
velocity (ms⁻¹)	0	5	10	15	15	15	0

b

time (s)	0	10	15	30	40	45	55
velocity (ms⁻¹)	0	20	20	10	10	30	30

c

time (s)	0	5	10	15	20	25	30
velocity (ms⁻¹)	30	30	20	10	0	10	20

d

time (s)	0	2	8	14	16	20	25
velocity (ms⁻¹)	0	5	5	12	15	15	15

e

time (s)	0	10	15	20	25	32	38
velocity (ms⁻¹)	0	8	8	0	12	23	30

Hint   Take care when extracting values from velocity–time graphs.

1 For each of the graphs below, calculate the total displacement of the object.

a
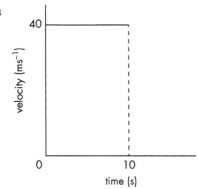

b
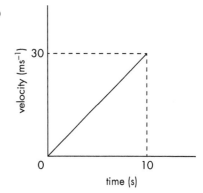

c
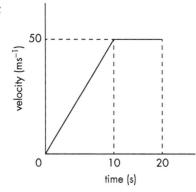

d
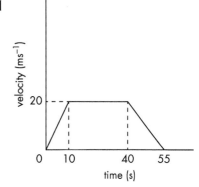

e
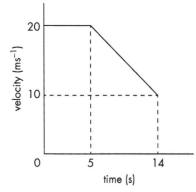

f
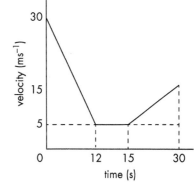

2 A car is travelling along a motorway at 22 ms⁻¹ when the driver sees queueing traffic ahead. A graph of the motion of the car is shown below.

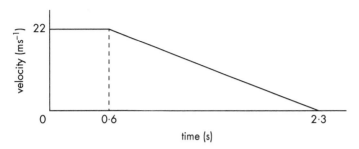

a At what time did the driver apply the brakes?

b Calculate the total stopping distance of the car from the moment the driver sees the queueing traffic.

3 The graph below shows the motion of a bouncing ball.

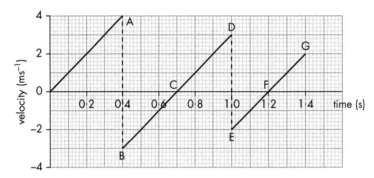

a Calculate how far the ball travels in the first 0.4 s.

> **Hint** A change in sign on the velocity–time graph (such as from positive to negative) indicates a change in direction of the object.

b Describe the motion of the ball at each point (0–A, A–B, etc.) indicated on the graph.

c Explain why the spikes on the graph decrease as time increases.

4 The following velocity–time graph represents the vertical motion of a ball.

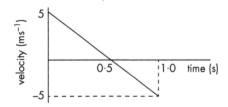

a Describe the motion of the ball during the first 0·5 seconds.

b Describe the motion of the ball between 0·5 seconds and 1 second.

c Calculate the acceleration of the ball during the first 0·5 seconds.

d Calculate the acceleration of the ball between 0·5 seconds and 1 second.

e What is the maximum height reached by the ball?

f Calculate the displacement of the ball after 1 second.

3 Acceleration

Exercise 3A Acceleration calculations

1. What is meant by an acceleration of 2 ms^{-2}?

2. A car is initially at rest. It accelerates at a constant rate of 5 ms^{-2} along a flat road. What speed will the car be travelling at after:

   a 1 s   **b** 2 s   **c** 5 s   **d** 7·5 s

3. A coin is dropped from the top of a tall building. It reaches a speed of 19·6 ms^{-1} after 2 seconds. What is the acceleration of the coin?

4. A sports car accelerates from rest to 27 ms^{-1} in 2·46 seconds. Calculate the acceleration.

5. A train increases speed from 12 ms^{-1} to 24 ms^{-1} in 6 seconds. Calculate the acceleration.

6. A plane is travelling along a runway at a constant speed of 15 ms^{-1}. It then accelerates at a constant rate for 18 seconds. The speed before take–off is 55 ms^{-1}. Calculate the acceleration.

7. A bus is travelling at 22 ms^{-1}.

   It takes 28 seconds to come to rest after its brakes are applied. Calculate the deceleration of the bus.

8. At one point during a 100 m race, Usain Bolt is travelling at a speed of 5·4 ms^{-1}. He accelerates at a constant rate of 3·1 ms^{-2} for 1·44 seconds. What is his final speed?

9. A motorbike accelerates at 1·2 ms^{-2}. It increases speed by 17·8 ms^{-1}. How long does the motorbike take to make this change in speed?

10. A skateboarder starts at rest from the top of a hill. At the bottom of the hill they reach a speed of 8 ms^{-1}. It takes 6 seconds to reach the bottom of the hill. Upon reaching the bottom of the hill, the skateboarder slows to a stop in a time of 12 seconds.

   a Calculate the acceleration of the skateboarder when travelling downhill.

   b Calculate the deceleration of the skateboarder as they come to rest.

11 A car is travelling at 14 ms⁻¹ when the driver sees a red light ahead. The driver applies the brakes and the car slows down with a constant acceleration of −2·5 ms⁻². Calculate the time taken for the car to slow to a stop.

12 A trolley with a card mounted on it is released from rest down a slope. The card passes through two light beams connected to two separate timers.

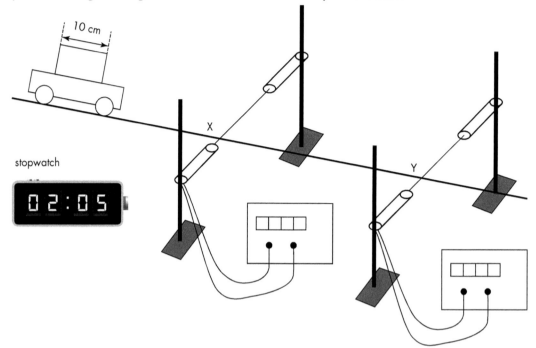

Describe how the apparatus can be used to measure the acceleration of the trolley.

Your description must include the following:

- measurements made
- equipment used
- any necessary calculations

1 Calculate the acceleration from the following velocity–time graphs:

a
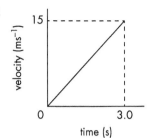

b
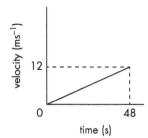

c
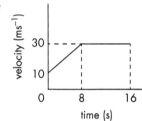

d
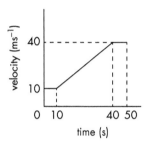

e
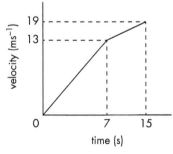

f
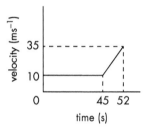

2 The graph shows the motion of a car over the first 15 seconds of its journey.

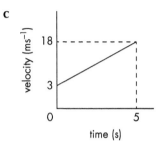

Use the information from the graph to calculate:

a the initial acceleration

b the acceleration between 7 and 15 seconds.

3 Use the velocity–time graph below to calculate the constant accelerations shown.

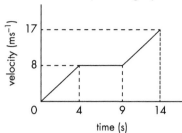

4 Look at the velocity–time graph below.

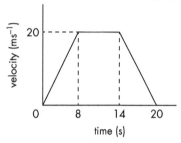

Calculate:

a the acceleration during the first 8 seconds

b the acceleration in the last 6 seconds

5 A lorry is approaching a roundabout and begins to decelerate. Use the information from the graph below to calculate the value of the deceleration at each stage.

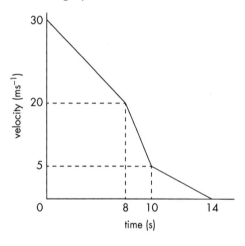

4 Newton's Laws

Exercise 4A Newton's First Law

1 State Newton's First Law.

2 Look at the free body diagrams shown below. Which ones show balanced forces?

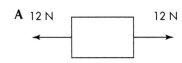
A 12 N   12 N

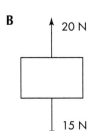
B   20 N   15 N

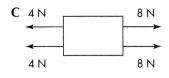
C 4 N   8 N
    4 N   8 N

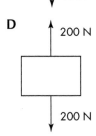
D   200 N   200 N

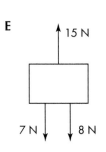
E   15 N
  7 N   8 N

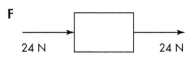
F   24 N   24 N

3 Each of the objects shown below is travelling at a constant speed. Identify the missing force in each case.

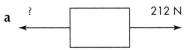
a ?   212 N

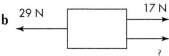
b 29 N   17 N   ?

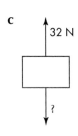
c 32 N   ?

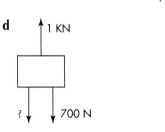
d 1 KN   ?   700 N

4   The downwards force on a shopping bag is 14 N. The bag is being held at a constant height. What is the magnitude of the upwards force?

5   An aeroplane is flying at a constant height of 11 000 m. The upwards force provided by the wings is 3 920 000 N. What is the size of the downwards force (weight)?

6   A helicopter has a weight of 7800 N downwards. The propellers provide an upward force of 7800 N. What can be said about the vertical position of the helicopter?

7   A cyclist is travelling along a straight, flat road at a constant speed of 10 ms⁻¹. The cyclist pedals with a force of 520 N. What is the magnitude of the frictional force?

8   The velocity-time graph shows the motion of a parachutist.

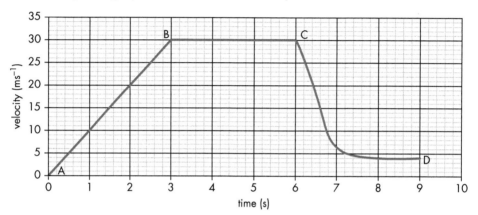

Identify the section(s) on the graph where the forces are balanced.

9   Car manufacturers state the maximum speed that a car can reach. They call this terminal velocity. Explain what is meant by terminal velocity in terms of the forces acting on the car.

10 A crash test dummy is used in a car to determine its safety rating. The car collides with a wall at high speed. The crash test dummy is not wearing a seatbelt. Describe the motion of the dummy using Newton's Laws of Motion.

Exercise 4B Newton's Second Law

1. State Newton's Second Law.

2. Look at the free body diagrams shown below. Calculate the unbalanced force and its direction in each case.

a

15 N
25 N

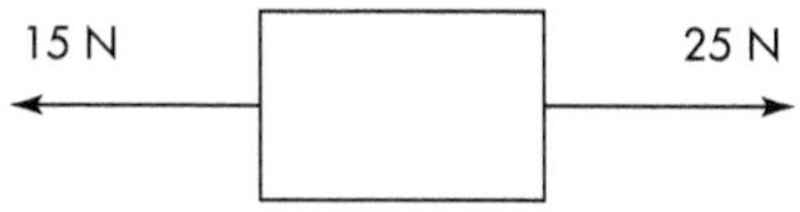

b

11 N
32 N

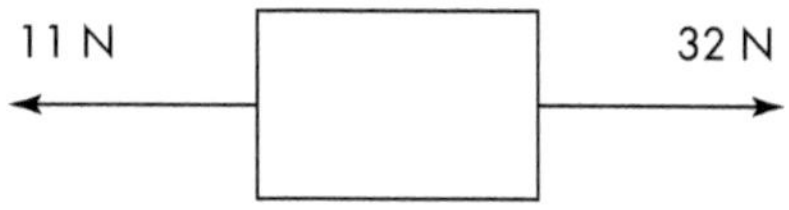

c

17 N
29 N

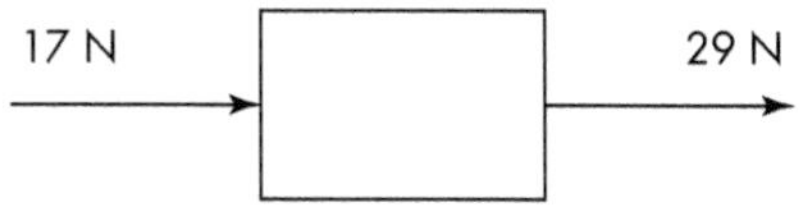

d

4 N
8 N
8 N

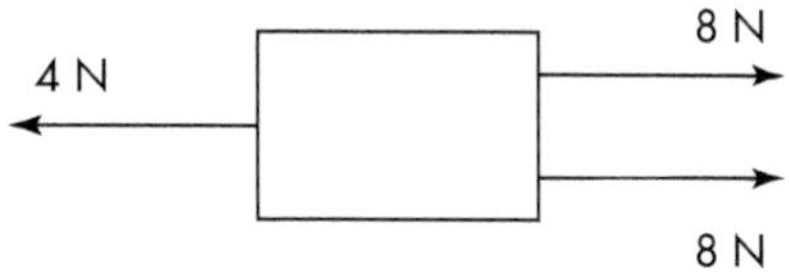

e

48 N
67 N

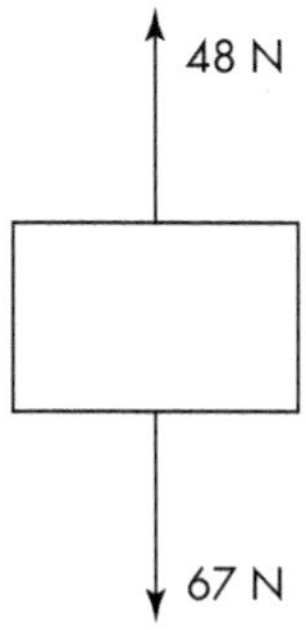

f

56 N
10 N
10 N
40 N

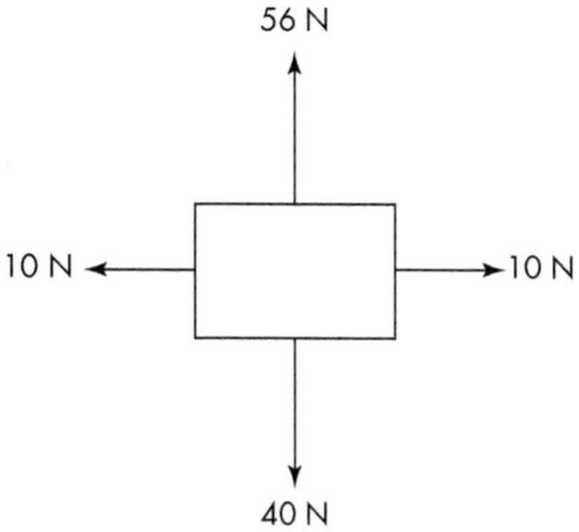

3. A car has a mass of 1000 kg. It accelerates at $1{\cdot}5\ ms^{-2}$. Calculate the unbalanced force.

4. A volleyball has a mass of 0·25 kg. After the ball is struck, it accelerates constantly at $8\ ms^{-2}$. Calculate the unbalanced force.

5. A force of 180 N is applied to a mass of 50 kg. Calculate its acceleration.

6. An unbalanced force of 142 N acting on a supermarket trolley provides it with an acceleration of $4{\cdot}2\ ms^{-2}$. Calculate the mass of the trolley.

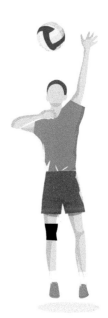

A block of mass 5 kg is pulled along a horizontal bench with a force of 20 N.

6 N ← [5 kg] → 20 N

A constant frictional force of 6 N acts on the block. Calculate the acceleration.

- If there is more than one force acting on an object then we must first calculate the unbalanced force. This can be thought of as the extra force going in a particular direction. In this case:

  $F = 20 - 6 = 14$ N

- This value can now be used in the equation for Newton's Second Law, $F = ma$:

$$a = \frac{F}{m}$$

$$a = \frac{14}{5}$$

$$a = 2.8 \text{ ms}^{-2}$$

7 Calculate the acceleration of each object shown.

a 25 N ← [20 kg] → 100 N

b 4500 N ← [280 kg] → 2250 N

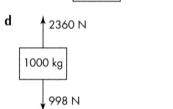

c 6.8 N ← [1.3 kg] → 8.4 N

d

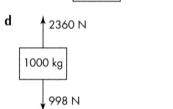

2360 N ↑ [1000 kg] ↓ 998 N

8 A car's engine produces a force of 17 000 N. The mass of the car is 1175 kg. The force of friction acting on the car is 6920 N.

a Calculate the unbalanced force acting on the car.

b Calculate the acceleration of the car.

9 A train is constantly accelerating at 5 ms^{-2}. The train has a mass of 146 000 kg. The force provided by the engine is 820 kN.

a Calculate the unbalanced force acting on the train.

b Calculate the force of friction.

10 A cyclist and their bike have a combined mass of 65 kg. The cyclist accelerates constantly from 4 ms^{-1} to 9.2 ms^{-1} in 14 seconds. The force of friction acting against the cyclist is a constant 68 N.

a Calculate the unbalanced force.

b Calculate the force produced by the cyclist.

1 Describe the difference between weight and mass.

> Hint   Values of gravitational field strength can be found in the data sheet.

2 Calculate the weight of each of the following objects on Earth.

  a 15 kg dog   **b** 65 kg girl   **c** 1125 kg car

  d 0·5 kg ball   **e** 200 g bag of sugar   **f** 4 mg ant

3 An astronaut has a mass of 89 kg.

  a Using the data below, calculate the weight of the astronaut on each of the planets.

Planet	Gravitational field strength (Nkg^{-1})
Earth	9·8
Jupiter	23
Neptune	11
Saturn	9·0
Mercury	3·7

  b How would the mass of the astronaut on Earth compare to the mass of the astronaut on Mars?

4 An object has a weight of 750 N. Calculate the mass of the object on:

  a Earth

  b Jupiter

  c the Moon

5 A 42·5 kg box is found to have a weight of 414 N. Calculate the value of the gravitational field strength.

6 A helicopter has a mass of 1170 kg. The helicopter accelerates upwards at 12 ms^{-2}. Calculate the unbalanced force.

7 A hot air balloon has a mass of 3250 kg. The passengers have a combined mass of 520 kg. If the unbalanced force acting upwards is 415 N, calculate the upwards acceleration.

8 A rocket has a mass of 3000 kg. The rocket engines produce a constant thrust force of 30 000 N. Calculate:

  a the weight of the rocket before it takes off from Earth

  b the unbalanced force

  c the initial acceleration of the rocket

9 A bodybuilder can exert a maximum upward force of 800 N. Assuming air resistance is negligible, calculate the maximum acceleration they can provide to a lifted mass of 65 kg.

10 A person is rescued from the ocean by a coastguard helicopter. The stretcher and person have a combined mass of 80 kg.

cable

a Calculate the combined weight of the person and the stretcher.

b The cable from the helicopter provides an upward force of 1250 N to lift the person and the stretcher. Calculate the acceleration of the person and the stretcher.

11 A balloon of mass 600 kg rises vertically from the ground. The graph below shows how the vertical speed of the balloon varies over time.

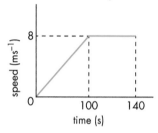

a Calculate the initial acceleration of the balloon.

b Calculate the weight of the balloon.

c Calculate the upward force acting on the balloon during the first 100 seconds of the flight.

12 A liquid-fuelled rocket takes off from the Moon vertically upwards. The engines produce a thrust of 120 000 N. The weight of the rocket is 40 000 N.

a Find the initial acceleration of the rocket.

b Assuming there is a constant thrust from the burning fuel, explain what happens to the acceleration of the rocket as it rises.

c Would this rocket take off on Earth? Justify your answer.

1 State Newton's Third Law.

2 Identify the Newton pair of forces in each of the situations below:

a

b

c

d

3 Two people are pulling on a rope in opposite directions. The action force is the hand pulling on the rope. Identify the reaction force.

4 A student makes the following statement: 'When I stand still on a solid floor, the only force on me is the force of Weight acting downwards'. Comment on why you think this statement is true or false.

5 Explain, using Newton's Third Law, how a rocket is able to take off from the surface of the Earth.

5 Energy

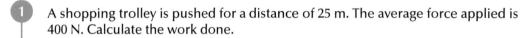

Exercise 5A Work done

1. A shopping trolley is pushed for a distance of 25 m. The average force applied is 400 N. Calculate the work done.

2. A lawnmower is pushed with an average force of 230 N over a distance of 42 m.

   Calculate the work done.

3. A car is brought to rest over a distance of 24 m. The frictional force is 980 N. Calculate the work done in bringing the car to rest.

4. A pram is pushed for a distance of 700 m. The total work done is 210 kJ. Calculate the average force applied.

5. A car tows a caravan for a distance of 12 km. The total work done is 30 MJ. Calculate the average force applied.

6. A tug boat is towing a small oil rig. The boat does 205 MJ of work by exerting an average force of 15 kN on the oil rig. Calculate the distance the oil rig is towed.

7. A golf club exerts a force of 370 kN on a golf ball. The work done on the ball is 21 kJ. Calculate the distance over which the force is applied.

8. A cross-country skier experiences a frictional force of 145 N.

   The unbalanced force is 325 N. If the distance travelled is 82 m, what is the work done by the skier?

9. A rock climber has a weight of 722 N. How much work is done when climbing a vertical distance of 67 m?

10 A person of mass 50 kg runs up a flight of stairs. The vertical height of the stairs is 8 m. Calculate:

a the weight of the person

b the work done

11 An exercise bike is shown below.

brake

wheel

a A person using the exercise bike pedals against frictional forces applied to the wheel by the brake. A frictional force of 200 N is applied at the edge of the wheel, which has a circumference of 1·2 m. How much work is done by friction in 600 turns of the wheel?

b The wheel is a solid aluminium disc that has a mass of 18·0 kg. Assuming all the friction is converted to heat in the disc, calculate the temperature rise after 600 turns.

Hint Use conservation of energy.

c Explain why the actual temperature rise of the disc is less than that calculated in **b**.

Exercise 5B Gravitational potential energy

1. What is meant by the term gravitational potential energy?

2. A mass of 25 kg is 10 m above the surface of the Earth. Calculate the gravitational potential energy.

3. A crane lifts a shipping container to a height of 80 m. The mass of the container is 24 000 kg. Calculate the gravitational potential energy gained by the container.

4. A climber has a mass of 76 kg.

   The climber gained 42 kJ of gravitational potential energy during a climb. Calculate the height.

5. The water in the reservoir of a hydroelectric power station has a mass of 30 000 kg. The water has a gravitational potential energy of 23 MJ. Calculate the height of the water in the reservoir.

6. The gravitational potential energy of a ball is 48 J. The height of the ball above the ground is 31 m. Calculate the mass of the ball.

7. A pole vaulter jumps a vertical height of 4·72 m. Gravitational potential energy of 3109 J is gained during the jump. Calculate the mass of the pole vaulter.

8. A space module takes off from the surface of a planet. The module has a mass of 356 kg. Once the module reaches a height of 45 m it has gained 37·8 kJ of gravitational potential energy. Calculate the gravitational field strength of the planet.

9. A coastguard helicopter lifts an injured sailor to safety. The sailor, of mass 84 kg, is lifted to a height of 39 m. Calculate:

   a the work done lifting the sailor

   b the gravitational potential energy gained by the sailor

10. An athlete completing an assault course climbs a vertical rope. The height of the rope climb is 17 m. The mass of the athlete is 58 kg. Calculate:

   a the weight of the athlete

   b the work done in climbing the rope

   c the gravitational potential energy gained by the athlete

1. What is meant by the term kinetic energy?

2. A car has a mass of 1100 kg and is travelling at 8 ms⁻¹. Calculate the kinetic energy of the car.

3. A plane has a mass of 9000 kg. It flies at a speed of 100 ms⁻¹. Calculate the kinetic energy of the plane.

4. A runner crosses the finishing line at a speed of 7 ms⁻¹. At this point the runner has a kinetic energy of 1862 J. Calculate the mass of the runner.

5. A remote-controlled car has a speed of 1·2 ms⁻¹. It has a kinetic energy of 0·216 J. Calculate the mass of the car.

6. An apple falling from a tree has a mass of 100 g. It has a kinetic energy of 5 J when it reaches the ground. Calculate the speed of the falling apple as it hits the ground.

7. A lorry of mass 2800 kg is travelling at a constant speed. It has a kinetic energy of 132 kJ. Calculate the speed of the lorry.

8. A motorbike has a mass of 250 kg and is travelling at 18 ms⁻¹.

A car has a mass of 950 kg and travels at 12·5 ms⁻¹.

Show by calculation which has the greater kinetic energy.

9. A car of mass 850 kg is travelling at a speed of 20 ms⁻¹. The brakes are applied and the car reaches a speed of 12 ms⁻¹. Calculate the change in the car's kinetic energy.

10. A rocket of mass 90 000 kg accelerates from rest at 5·2 ms⁻² for 15 seconds.

    a What speed is the rocket travelling at after 15 seconds?

    b What is the kinetic energy gained by the rocket?

Exercise 5D Conservation of energy

 State and explain the principle of conservation of energy.

 A large boulder of mass 8 kg falls from a cliff that is 20 m high.

  a Calculate the gravitational potential energy lost by the boulder.

  b Calculate the kinetic energy gained by the boulder, assuming no energy is lost to heat or sound.

  c Calculate the speed of the boulder when it hits the ground.

A diver of mass 70 kg dives from a high board which is 15 m above the water. Assuming air friction is negligible, calculate the speed of the diver just before they hit the surface of the water.

- First calculate the gravitational potential energy the diver has at a height of 15 m:

$E_p = mgh$

$E_p = 70 \times 9 \cdot 8 \times 15$

$E_p = 10\,290 \text{ J}$

- Because air friction is negligible, we can assume that all gravitational potential energy is converted to kinetic energy. Therefore, $E_p = E_k$

$$E_k = \frac{1}{2}mv^2$$

- Rearrange for v:

$$v^2 = \frac{E_k}{\frac{1}{2}m}$$

- Apply conservation of energy:

$$v^2 = \frac{10\,290}{\frac{1}{2} \times 70}$$

$v^2 = 294$

$v = \sqrt{294}$

$v = 17 \text{ ms}^{-1}$

3 A ball of mass 250 g is dropped from a height of 12 m. Calculate the speed of the ball when it hits the ground.

4 An astronaut on the Moon drops a hammer of mass 0·75 kg from a height of 1·32 m. Calculate the speed of the hammer when it lands on the surface of the Moon.

5 A box of mass 0·65 kg falls off a table and hits the ground at 4 ms⁻¹. Calculate the height of the table.

6 A helicopter drops a 43·5 kg crate of food supplies. The crate hits the ground at 23·8 ms⁻¹. Calculate the height from which the crate was dropped.

7 A car on a ramp is raised to a height of 2·8 m. The mass of the car is 995 kg.

  a Calculate the gravitational potential energy gained by the car.

  b How much work is done in raising the car to this height?

  c Calculate the force supplied to the ramp.

8 A 40-g golf ball is launched into the air with an initial vertical speed of 12 ms⁻¹. Calculate the maximum height the ball will reach.

9 A skateboarder and skateboard have a combined mass of 66·75 kg. The skateboarder has a speed of 7 ms⁻¹ at point A on the slope. The skateboarder comes to rest at point B. Assume that frictional effects are negligible.

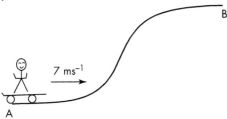

   a Calculate the kinetic energy of the skateboarder at point A.

   b What is the potential energy at point B?

   c Calculate the height of the slope.

10 A skier travels down a 100 m slope from rest before ascending a smaller slope of height 70 m. Assuming friction is negligible, what is the speed of the skier at point Y?

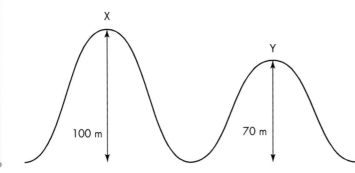

6 Projectile motion

Exercise 6A Projectile motion

1 Describe what is meant by projectile motion.

2 A ball is launched horizontally from a cliff at a speed of 15 ms⁻¹. It hits the surface of the water after 3 seconds.

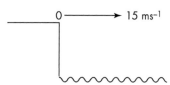

 a Calculate the horizontal distance travelled by the ball.

 b Calculate the vertical velocity of the ball just before it hits the water.

3 A golfer hits a shot from the tee at the top of a slope. The golf ball is struck horizontally at a speed of 35 ms⁻¹. It lands on the fairway below 4·2 seconds later.

 Calculate:

 a the horizontal distance travelled by the ball

 b the final vertical speed of the ball just before it lands on the fairway

4 Use the graphs below to calculate the horizontal displacement of the projectile.

 a 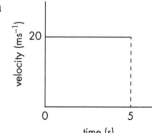   **b** 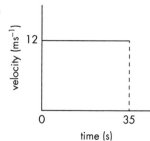   **c**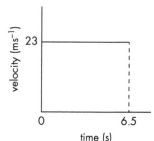

5 Use the graphs below to calculate the vertical displacement of the projectile.

 a 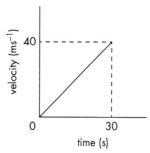   **b** 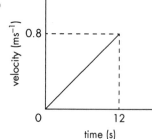   **c**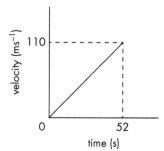

6 A parcel is dropped from a plane travelling in a straight line horizontally at a constant speed of 80 ms^{-1}. The parcel hits the ground 25 seconds later. The effects of air resistance can be ignored.

   a Calculate the horizontal distance travelled by the parcel.

   b What is the horizontal displacement of the parcel when it hits the ground, relative to the plane?

   c Calculate the vertical velocity of the parcel as it hits the ground.

   d What is the height of the plane when the parcel was dropped?

7 A stone dropped from rest down a well takes 1·6 seconds to hit the bottom. How deep is the well?

8 A diver jumps horizontally from a high diving platform at a speed of 5 ms^{-1}. The diver hits the surface of the water after 1·4 seconds.

   a Calculate the final vertical speed of the diver.

   b Sketch a velocity-time graph of the horizontal motion of the diver. Include numerical values on both axes.

   c Calculate the horizontal distance travelled by the diver.

   d Sketch a velocity-time graph of the vertical motion of the diver. Include numerical values on both axes.

   e Calculate the height of the diving platform above the surface of the water.

9 A ball rolls horizontally off a table at 2 ms^{-1} and hits the ground after 0·6 seconds.

   a Calculate the horizontal distance travelled by the ball.

   b Sketch a velocity-time graph of the horizontal motion of the ball. Include numerical values on both axes.

   c Calculate the vertical velocity of the ball as it hits the ground.

   d Sketch a velocity-time graph of the vertical motion of the ball. Include numerical values on both axes.

   e Calculate the height of the table.

10 By considering projectile motion, explain why a satellite stays in orbit.

7 Space exploration

Exercise 7A Space exploration

1. Give a definition of the following terms:
   a planet
   b dwarf planet
   c moon
   d Sun
   e asteroid
   f solar system
   g star
   h exoplanet
   i galaxy
   j Universe

2. List three benefits of satellites.

3. A GPS satellite orbits the Earth at an altitude of 2000 km. Calculate the time taken for the signal to travel from the satellite to a GPS receiver on Earth.

4. A satellite is in orbit 30 000 km above the Earth's surface. A signal of frequency 8 GHz is used to transmit the signal from a ground station to the satellite. Calculate:
   a the wavelength of the signal
   b the time taken for the signal to travel from the ground station to the satellite

5. a What is meant by the period of a satellite?
   b What happens to the period of a satellite when orbital altitude is increased?

6. a What is a geostationary satellite?
   b What is the period of a geostationary satellite?
   c What is the orbital altitude of a geostationary satellite?

7. Suggest a solution to the following challenges of space travel:
   a travelling large distances
   b maintaining sufficient energy to operate life-support systems

8. What are the risks associated with manned space exploration?

8 Cosmology

Exercise 8A Light years

 1 What is the definition of a light year?

 2 Convert the following distances from light years into metres:

    a 1 light year

    b 4 light years

    c 27 light years

    d 0·28 light years

 3 Convert the following distances from metres to light years:

    a 3×10^{10} m

    b $1{\cdot}9 \times 10^{19}$ m

    c $1{\cdot}5 \times 10^{11}$ m

    d $3{\cdot}83 \times 10^{17}$ m

4 It takes light from the Sun 8 minutes to reach the Earth. What is the distance from the Earth to the Sun, in metres?

A black hole, Sagittarius A* is found to be approximately 26 000 light years from Earth. Calculate the distance to Sagittarius A* in metres.

- Use the equation d = vt, where v is the speed of light and t is the time in seconds:

  $d = vt$

- This means we must convert 26 000 years into seconds. Start by converting years to days:

  $26\,000 \times 365$

- Next, convert days to hours:

  $26\,000 \times 365 \times 24$

- Hours to minutes:

  $26\,000 \times 365 \times 24 \times 60$

- And finally, hours to seconds:

  $26\,000 \times 365 \times 24 \times 60 \times 60$

- We can now perform the calculation:

  $d = 3 \times 10^{8} \times (26\,000 \times 365 \times 24 \times 60 \times 60)$

  $= 2{\cdot}46 \times 10^{20}$ m

 5 The Large Magellanic Cloud is a satellite galaxy of the Milky Way. It is a distance of 158 200 light years away from Earth. Calculate the distance to the Large Magellanic Cloud in metres.

6 The nearest star outside of our solar system is Proxima Centauri. It takes light from Proxima Centauri 4·3 years to reach Earth. What is the distance between Earth and Proxima Centauri, in metres?

7 The Milky Way is approximately 100 000 light years across. What is this distance in kilometres?

8 Andromeda is the nearest galaxy to the Milky Way. The distance to Andromeda is $2\cdot4 \times 10^{22}$ m. How long does it take light from Andromeda to reach our galaxy?

9 The mean distance from Earth to Neptune is $4\cdot5 \times 10^9$ km.

  a How long does it take light from Neptune to reach Earth?

  b What percentage of a light year is this distance?

10 The dwarf planet Pluto is at a mean distance of $7\cdot5 \times 10^9$ km from Earth.

  a How long does it take the light from Pluto to reach Earth?

  b How long would it take to reach Pluto travelling at a constant speed of 1000 ms⁻¹?

  c How long would it take to reach Pluto travelling at 27 ms⁻¹ (approximately 60 mph)?

Exercise 8B The Universe and spectra

1 Describe the Big Bang theory of the origin of the Universe.

2 What is the approximate estimated age of the Universe?

3 The whole of the electromagnetic spectrum is used to study astronomical objects. Why is it useful to study objects using signals other than the visible spectrum?

4 Identify the type of spectra shown below:

a

b

5 The diagram below shows the line spectrum for elements X, Y and Z together with the spectrum from a star.

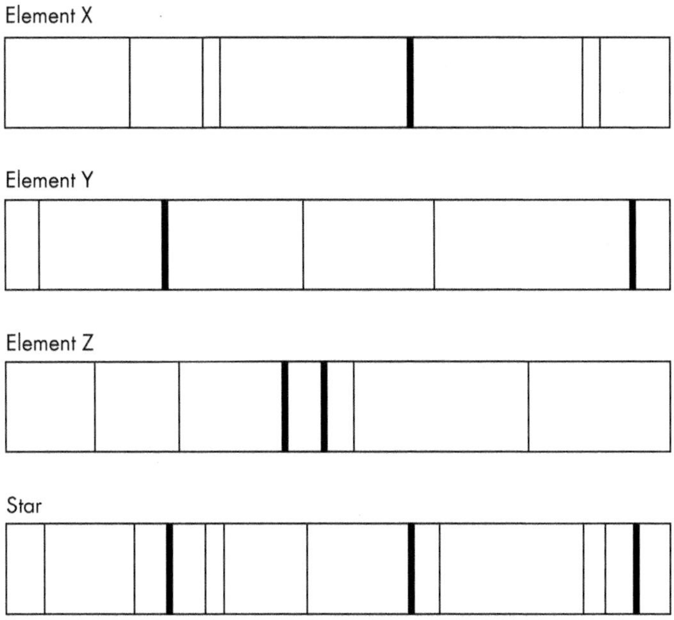

Element X

Element Y

Element Z

Star

a Is the element X present in the star spectrum?

b Is the element Y present in the star spectrum?

c Is the element Z present in the star spectrum?

6 What elements does the unknown sample contain?

Unknown sample

Element A

Element B

Element C

Element D

7 The line spectrum for a star is shown along with the line spectra for calcium, helium, hydrogen and sodium.

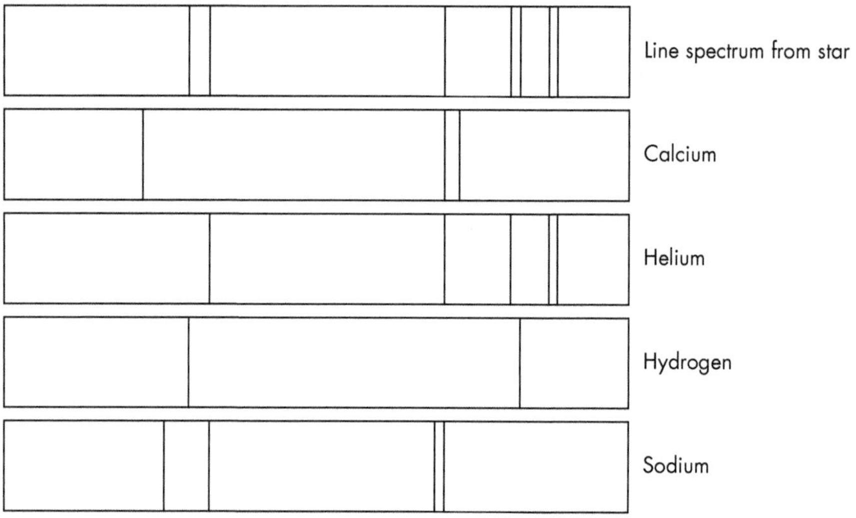

Line spectrum from star

Calcium

Helium

Hydrogen

Sodium

Use this information to find the elements present in the star's atmosphere.

8 The line spectrum of a star is shown along with the line spectra for hydrogen, helium, sodium and mercury.

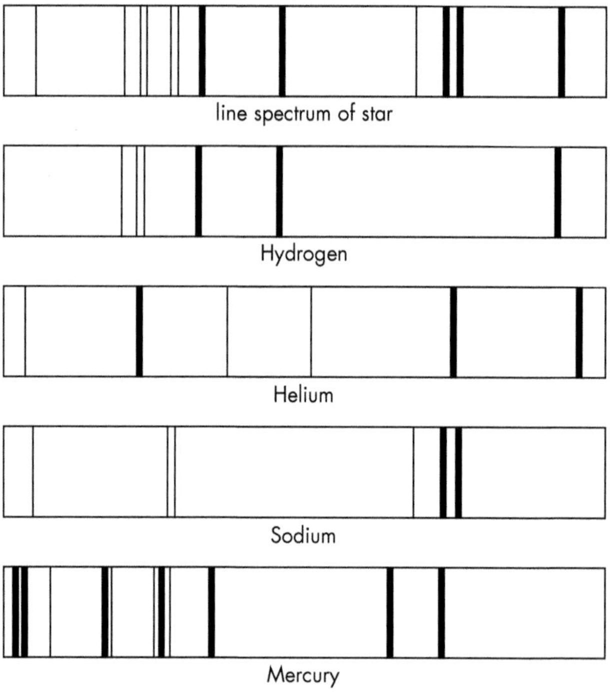

line spectrum of star

Hydrogen

Helium

Sodium

Mercury

Identify the element(s) present in the line spectrum of the star.

9

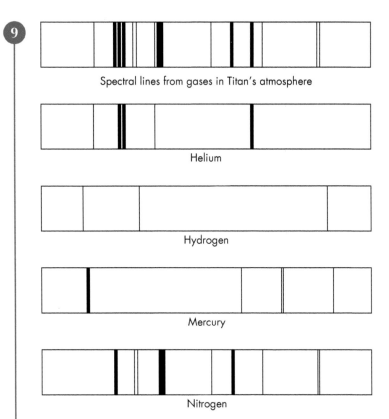

Spectral lines from gases in Titan's atmosphere

Helium

Hydrogen

Mercury

Nitrogen

Use the spectral lines of the elements to determine which elements are present in the atmosphere of Titan.

10 The spectral lines for elements W, X, Y and Z are shown below.

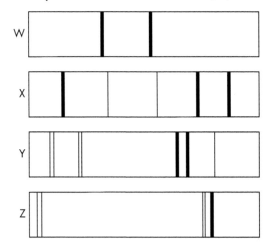

W

X

Y

Z

The line spectrum of a star is found to contain elements W and Y. Draw the line spectrum for the star.

9 Electrical charge carriers

Exercise 9A Current, charge and time

1 State the definition of electrical current.

2 What charge will have passed through a metal wire when there is a current of 10A for 20 seconds?

3 A laptop computer operates at a constant current of 0·5 A.
The laptop is switched on for 4 hours. Calculate the charge.

> Hint Time must be converted to seconds.

4 212 C of charge is drawn from a battery in 2 minutes. Calculate the current leaving the battery.

5 A Playstation console is used for 45 minutes. During this time, 2320 C of charge flows through the console. Calculate the electric current.

6 An LED is operating at a maximum current of 3·5 mA. How long will it take for 0·5 C of charge to pass through the LED?

7 A power supply produces 0·2 mA of current. How long will it take the power supply to produce a charge of 1·5 C?

8 The current in a hairdryer is 10A.
The hairdryer is used for 8 minutes.

a Calculate the charge.

b If the charge on an electron is $1·6 \times 10^{-19}$ C, calculate the number of electrons for the charge calculated in **a**.

Exercise 9B Alternating and direct current

1 Explain, in terms of electron flow, the difference between alternating and direct current.

2 Give an example of a source of:

a alternating current

b direct current

3 From the two images below, decide which shows:

a a.c.

b d.c.

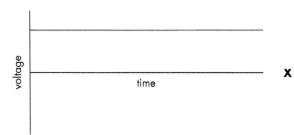

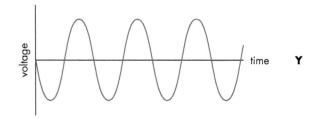

10 Potential difference (voltage)

Exercise 10A Potential difference

1 What is the definition of the potential difference of a power supply?

2 What are the two types of electric charge?

3 Describe what happens when:

a two like charges are close together

b two unlike charges are close together

4 There is a uniform electric field between plates Q and R. A particle, P, passes through the field as shown.

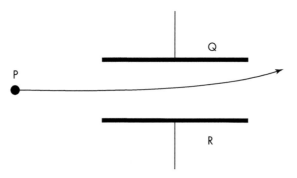

If the particle is negatively charged, what is the charge on:

a Plate Q?

b Plate R?

5 The diagram below shows protons, neutrons and electrons in an electric field.

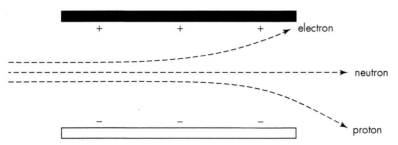

Redraw the diagram to show how alpha particles, beta particles and gamma rays would travel through the electric field.

11 Ohm's Law

Exercise 11A Ohm's Law

1 A 150 Ω lamp has a current of 1·35 A. Calculate the voltage of the lamp.

2 An electric drill is connected to a 230 V mains supply. The drill has a resistance of 94 kΩ. Calculate the current.

3 An iron has a voltage of 210 V and a current rating of 12·2 A. Calculate the resistance of the iron.

4 An engineer sets up the circuit below to verify Ohm's Law.

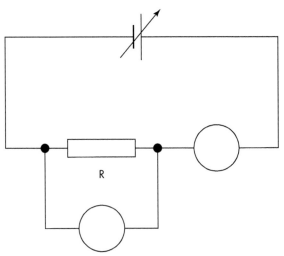

a Copy and complete the circuit, indicating which meter is the ammeter and which meter is the voltmeter.

b Describe how the circuit above could be used to verify Ohm's Law.

5 A student measures the following readings of voltage and current for a circuit containing a single resistor:

Voltage (V)	2·0	4·5	6·0	7·5	10
Current (mA)	40·0	90·0	119	149	200

a Plot a graph of voltage against current and use the graph to derive the relationship between voltage and current.

> **Hint**   Label both axes on your graph with the quantity and its units.

b Using your graph, or otherwise, find the resistance of the resistor.

6 What is the relationship between the temperature and the resistance of a conductor?

7 Find the voltage across each resistor shown.

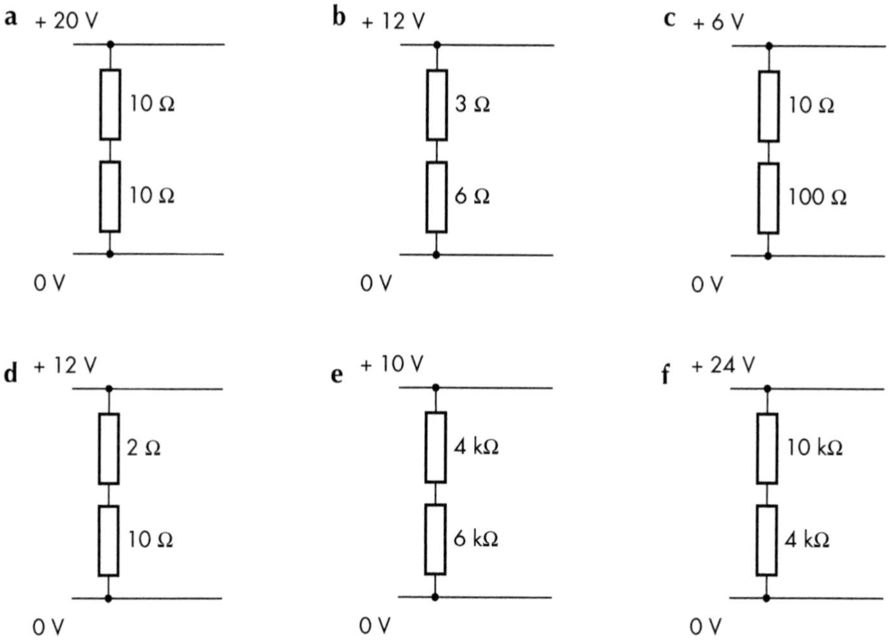

a + 20 V

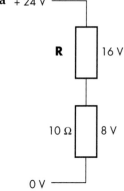

10 Ω

10 Ω

0 V

b + 12 V

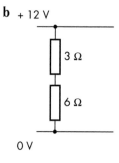

3 Ω

6 Ω

0 V

c + 6 V

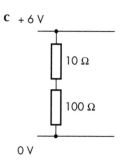

10 Ω

100 Ω

0 V

d + 12 V

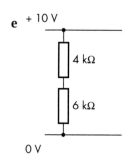

2 Ω

10 Ω

0 V

e + 10 V

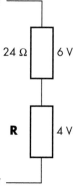

4 kΩ

6 kΩ

0 V

f + 24 V

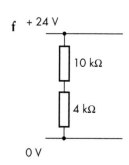

10 kΩ

4 kΩ

0 V

8 Calculate the missing values in each of the diagrams below.

a + 24 V

R   16 V

10 Ω   8 V

0 V

b + 10 V

24 Ω   6 V

R   4 V

0 V

c + 20 V

6 kΩ   **V**

2 kΩ   5 V

0 V

d + **V**$_s$

60 Ω   25 V

30 Ω   **V**

0 V

e + **V**$_s$

12 kΩ   10·8 V

20 kΩ   **V**

0 V

f + **V**$_s$

3·8 kΩ   **V**

1·7 kΩ   3·2 V

0 V

Exercise 12A Series circuits

1 Two identical bulbs are connected as shown below.

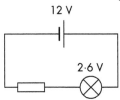

5 V   5 V

What is the value of the supply voltage?

2 A circuit with a lamp and a resistor in series is shown below.

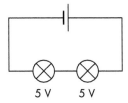

12 V

2·6 V

a What is the voltage across the resistor?

b If the current in the resistor is 1 A, what is the current in the lamp?

3 Three identical resistors are connected to a 15 V supply. Calculate the voltage across each resistor.

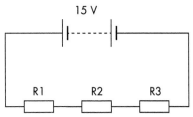

15 V

R1   R2   R3

4 Calculate the missing currents and voltages in each of the circuits below.

a

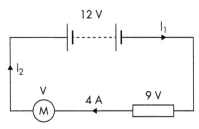

12 V

I_1

I_2

V

M

4 A   9 V

b

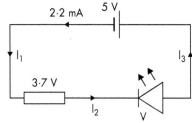

2·2 mA   5 V

I_1 I_3

3·7 V

I_2 V

c

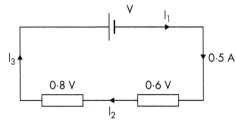

V I_1

I_3 0·5 A

0·8 V   0·6 V

I_2

Exercise 12B Parallel circuits

1 Calculate the missing current values in the circuits below.

a

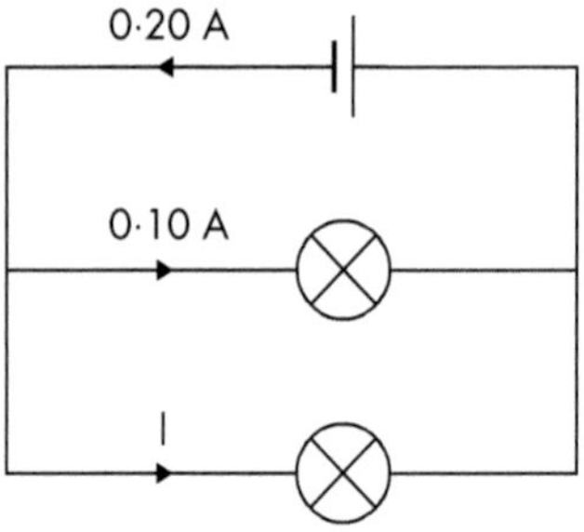

0·20 A

0·10 A

I

b

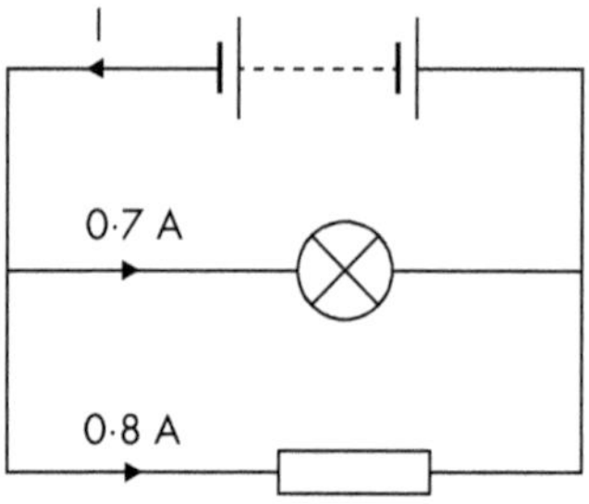

I

0·7 A

0·8 A

2 Two identical resistors are connected as shown below. Calculate the missing current values.

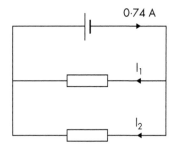

0·74 A

I_1

I_2

3 State the voltage across the lamp in each of the circuits shown below.

a

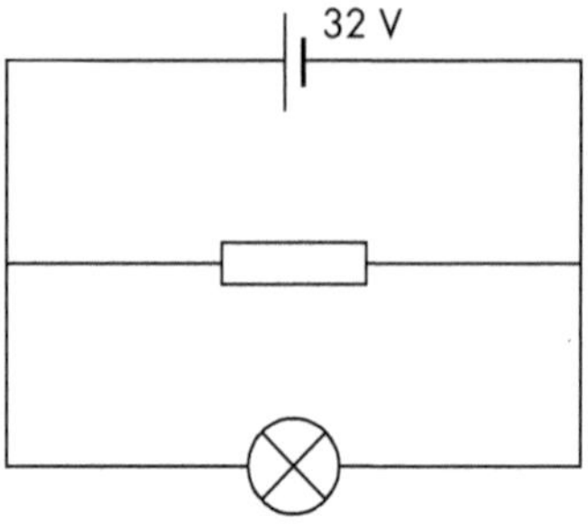

32 V

b

12·8 V

4 Calculate the missing current and voltage values in each of the circuits below.

a

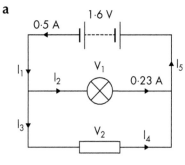

0·5 A   1·6 V

I_1 I_2 V_1 I_5

0·23 A

I_3 V_2 I_4

b

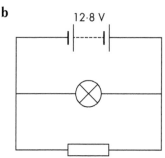

12 V   0·3 A

V_1 I_1

120 Ω

I_5 V_2 I_2

120 Ω

I_4 V_3 I_3

120 Ω

c

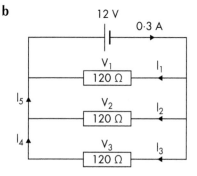

12 V

1·2 A

I_4 V_1 I_1

160 Ω

I_3 V_2 I_2

160 Ω

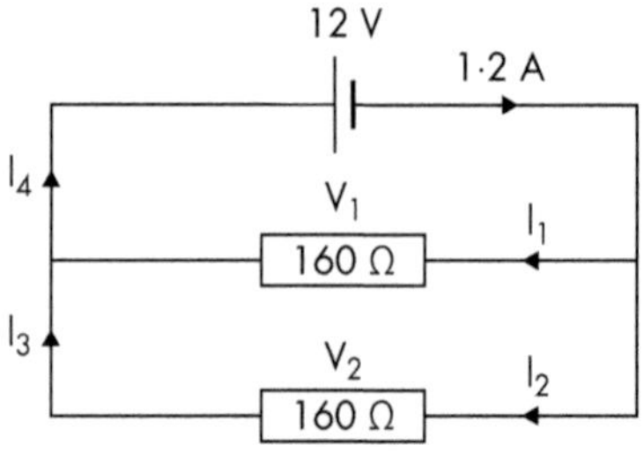

5 What are the advantages of using parallel circuits instead of series circuits?

Exercise 12C Circuit symbols

1 For each of the symbols shown below, identify the component and describe its function:

a ——————| |————————

b ——| |- - - - - -| |——

c ————⊗————

d ——o ⁄ o——

e ——[ ]————

f ——(V)————

g ——(A)————

h

i ——(M)————

j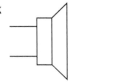

k

l

m ——[ ]————

n ——▷|————

o ——| |————

p ——[/]————

q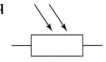

r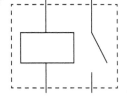

2 **a** Draw the symbol for an NPN transistor.
   b Draw the symbol for an n-channel enhancement MOSFET.
   c What is the function of a transistor in a circuit?

Exercise 12D Resistors in series and parallel circuits

1 What happens to the total resistance of a circuit when more resistors are added in series?

2 Calculate the total resistance of the following circuits.

a

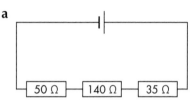

b

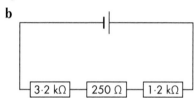

3 The total resistance of the circuit shown is 3·6 kΩ. Calculate the resistance of resistor R.

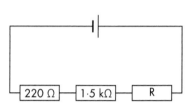

4 What happens to the total resistance of a circuit as more resistors are added in parallel?

5 Calculate the total resistance of the following circuits.

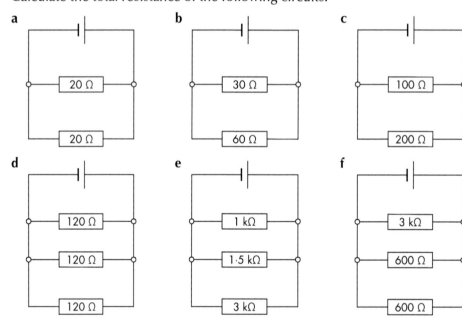

Find the total resistance of the circuit shown below.

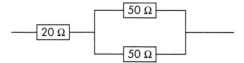

The first step is to make the circuit as simple as possible. The best way to do this here is to calculate the resistance of the parallel branch.

$$\frac{1}{R_p} = \frac{1}{R_1} + \frac{1}{R_2}$$

$$\frac{1}{R_p} = \frac{1}{50} + \frac{1}{50}$$

$$\frac{1}{R_p} = \frac{2}{50}$$

$R_p = 25\ \Omega$

The circuit now looks like this:

—[20 Ω]——[25 Ω]—

This is a series circuit so the total resistance is calculated from the sum of the resistance values:

$R_T = R_1 + R_2$

$R_T = 20 + 25$

$R_T = 45\ \Omega$

6 Find the total resistance between points X and Y in each of the circuits below.

a

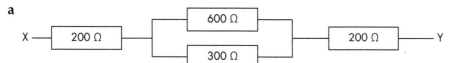

b

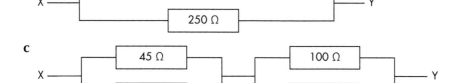

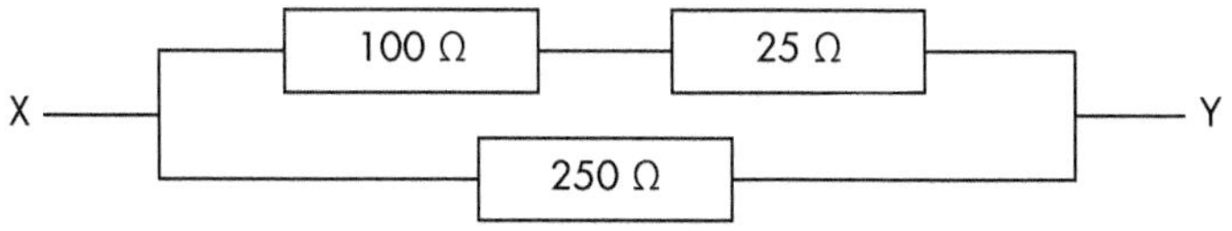

c

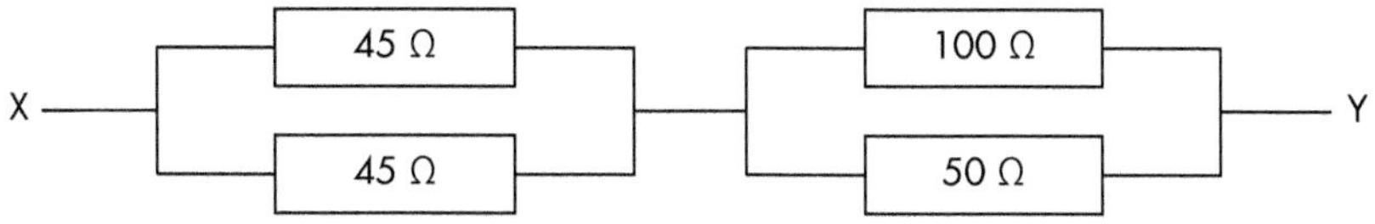

X —[45 Ω][45 Ω]——[100 Ω][50 Ω]— Y

Exercise 12E Complex circuits

1 Find the voltmeter reading in the circuit below.

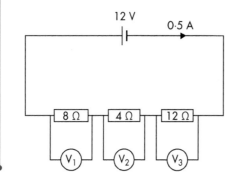

A student sets up the following circuit.

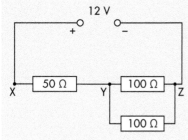

a Find the total resistance between points Y and Z.

- Resistors are in parallel so we must use:

$$\frac{1}{R_T} = \frac{1}{R_1} + \frac{1}{R_2}$$

$$\frac{1}{R_T} = \frac{1}{100} + \frac{1}{100}$$

$$\frac{1}{R_T} = \frac{2}{100}$$

$$R_T = 50\ \Omega$$

b i Find the potential difference across X and Y.

- First calculate the total resistance of the circuit. The 50 Ω calculated in **a** is in series with the 50 Ω resistor:

$$R_T = R_1 + R_2$$
$$R_T = 50 + 50$$
$$R_T = 100\ \Omega$$

- We can now apply Ohm's Law to find the current in the circuit, being careful to use the **total** resistance and the **supply** voltage:

$$V = IR$$

$$I = \frac{V}{R}$$

$$I = \frac{12}{100}$$

$$I = 0.12\ \text{A}$$

- Now calculate the voltage across X and Y:

$V = IR$

$V = 0.12 \times 50$

$V = 6\ V$

ii Find the potential difference across Y and Z.

$V_{YZ} = V_S - V_{XY}$

$V_{YZ} = 12 - 6$

$V_{YZ} = 6\ V$

2 For the circuit shown below, find:

a the current leaving the power supply

b the current in each resistor

c the voltage across each resistor

> **Hint** When working with complex circuits, try and simplify the circuit as much as possible. Resolving the resistors into one single resistor in series is often a useful approach.

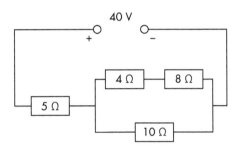

3 Calculate the current in the circuit below.

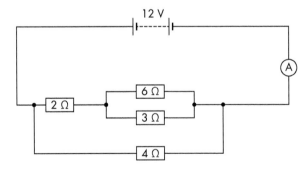

Exercise 12F Transistor switching circuits

1 The temperature of the water in a dishwasher is monitored using the circuit below.

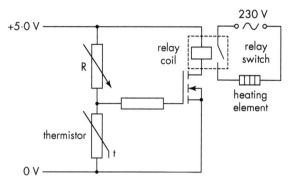

The heating element switches off when the temperature of the water is too high. Explain how the circuit operates to switch off the heating element.

2 The light level in a hotel room is monitored using the circuit below.

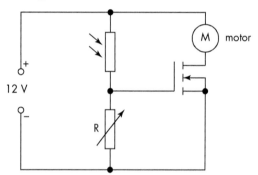

When the light level becomes too high the motor switches on, closing the blinds in the room. Explain how this circuit works to close the blinds.

3 An engineer designs the circuit shown below to switch on an LED when it is dark.

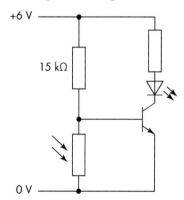

The table below shows the resistance of the LDR in dark and light conditions.

Light conditions	Resistance (kΩ)
Dark	9
Light	0·6

  a Calculate the voltage across the LDR when it is light.

  b Calculate the voltage across the LDR in dark conditions.

  c Explain how the circuit works to switch on the LED.

13 Electrical power

Exercise 13A Power, energy and time

1 Give the definition of power in terms of electrical energy and time.

2 Which fuse value should be used for an appliance with:

  a a power rating of less than 720 W?

  b a power rating of greater than 720 W?

3 What is the power rating of a heater that uses 3240 kJ of energy in 30 minutes?

4 A 2 kW kettle is left on for 110 seconds. Calculate the electrical energy transformed in this time.

5 How long will it take a 3500 W electric motor to convert 7890 J of electrical energy?

6 An electric heater converts 528 000 J of electrical to heat energy in a time of 4 minutes. Calculate the power rating of the electric heater.

7 A television with a power rating of 180 W is used for 3 hours. Calculate the energy used.

8 An LCD television has a power consumption of 121 W when switched on and 1·6 W when in standby mode.

  a Calculate the energy used when the television is switched on for 1 hour and 45 minutes.

  b How much energy is used when the television is in standby mode for 8 hours?

9 The power rating of three games consoles whilst playing a video game is given in the table below.

Games console	Power rating (W)
Console 1	137
Console 2	112
Console 3	18

  a Calculate the amount of electrical energy used by Console 1 in 45 minutes.

  b Calculate the time taken for Console 2 to use 202 kJ of energy.

  c Calculate the difference in the energy used by Console 1 and Console 3 in 1 hour.

10 A number of household appliances are used for a period of 2 hours. The appliances and their power ratings are listed below.

Appliance	Power rating (W)
Energy-saving lamp	60
LED television	58
Fridge	160
Halogen lamp	100

There are three energy-saving lamps, two halogen lamps, one LED television and one fridge. Calculate the total energy used by these appliances.

1 An electric fire is connected to a 230 V mains supply. The current in the element of the fire is 5 A. Calculate the power rating of the electric fire.

2 Calculate the current in a television of power rating 120 W and voltage 240 V.

3 What is the voltage of a 930 W hairdryer if the current is 4 A?

4 An iron has a power rating of 1·1 kW and a current of 4·3 A. Calculate the voltage.

5 A 118 V bulb has a current of 680 mA. Calculate the power rating of the bulb.

6 A 2·2 kW heater is connected to a 230 V mains supply. Calculate which fuse should be used in the heater: 3 A, 5 A or 13 A?

7 Three components, X, Y and Z, are tested, and the results are shown in table below.

Component	Voltage (V)	Current (mA)
X	12	250
Y	3·5	440
Z	6·0	565

Calculate the power rating for each component.

8 A car headlamp operates at 12 V with a current of 2·8 A.

Calculate:

a the power rating of the lamp

b the energy used when both front headlamps are switched on for 20 minutes

Exercise 13C Power, current and resistance

1 A loudspeaker has a resistance of 9 Ω. The current in the loudspeaker is 5·5 A. Calculate the power of the loudspeaker.

2 A 60 W light bulb has a current of 5 mA. Calculate the resistance.

3 A 44 Ω resistor is connected to a 5 V battery. Calculate the power rating of the resistor.

4 A 1·8 kW kettle has a resistance of 18 Ω. Calculate the current.

5 The current in a 10 MΩ resistor is 6 μA. Calculate the power rating of the resistor.

6 In the circuit shown, calculate:

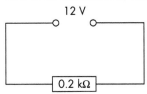

   a the current in the circuit

   b the power dissipated in the resistor

7 In the circuit shown, calculate:

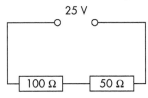

   a the total resistance of the circuit

   b the current in the circuit

   c the power rating of each resistor

8 In the circuit shown, calculate:

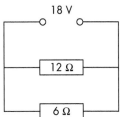

   a the total resistance

   b the current drawn from the battery

   c the current in each of the resistors

   d the power rating of each resistor

Exercise 13D Power, voltage and resistance

1 What is the resistance of a 120 W bulb connected to a 230 V mains supply?

2 A television has a resistance of 520 Ω and is connected to a voltage supply of 230 V. Calculate the power rating of the television.

3 Calculate the voltage of a buzzer which has a power rating of 9 W and a resistance of 220 Ω.

4 What is the resistance of a 60 W laptop operating at a voltage of 12 V?

5 Calculate the power rating of a smartphone charger that has a resistance of 5·5 Ω and a voltage of 22 V.

6 Calculate the voltage needed to operate two 100 W lamps connected in series if each bulb has a resistance of 280 Ω.

7 A washing machine has a current of 10 A and a resistance of 22 Ω.
Calculate:

a the power rating of the washing machine

b the electrical energy transformed in 1 hour 15 minutes

8 A 1300 W dishwasher is connected to a 230 V mains supply.
Calculate:

a the resistance of the dishwasher

b the current in the dishwasher

c the energy used by the dishwasher during a 40 minute cleaning cycle

14 Specific heat capacity

Exercise 14A Specific heat capacity

1 Explain what is meant by the specific heat capacity of a substance.

2 4 kg of water is heated from 20 °C to 30 °C. The specific heat capacity of water is 4180 J kg^{-1} °C^{-1}. Calculate the heat energy required for this change in temperature.

3 Calculate the heat energy required to change the temperature of 15 kg of aluminium from 10 °C to 70 °C.

4 Calculate the specific heat capacity of a material that requires 34 500 J of heat energy to change the temperature of 1 kg of the material by 35 °C.

5 2240 kJ of heat energy is needed for 80 kg of a substance to change temperature from 12 °C to 47 °C. Calculate the specific heat capacity of the substance.

6 Calculate the mass of copper if 30 000 J produces a temperature rise of 22 °C.

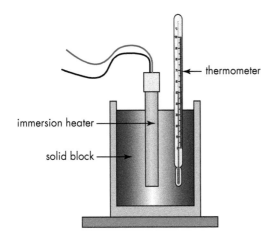

thermometer

immersion heater

solid block

7 A block of lead is heated from 20 °C to 33 °C. It takes 7·2 kJ of heat energy to cause this temperature change. Calculate the mass of the lead block.

8 300 g of alcohol absorbs 1340 J of heat energy. Calculate the change in temperature of the alcohol.

9 A kettle provides 199 kJ of heat energy to 540 g of water. The initial temperature of the water is 11 °C. Calculate the final temperature of the water.

10 A 1400 W kettle heats 1 kg of water from 38 °C to 92 °C. Calculate the time taken for this increase in temperature.

11 A 200 W heater is used to supply heat energy to a 3 kg block of copper. The heater is switched on for 8 minutes. Calculate the change in temperature of the copper block.

12 A piece of iron of mass 2 kg is dropped onto a surface without rebounding resulting in a temperature rise of 1 °C. Calculate the speed with which the iron hits the surface.

> **Hint** Use conservation of energy.

15 Specific latent heat

Exercise 15A Specific latent heat

1 a Explain what is meant by the latent heat of fusion of a substance.

   b Explain what is meant by the latent heat of vaporisation of a substance.

2 Calculate the heat energy required to change 1 kg of water to steam at 100 °C. The latent heat of vaporisation of water is $22 \cdot 6 \times 10^5$ J kg^{-1}.

3 Calculate the heat energy released when 0·8 kg of water freezes to ice at 0 °C.

4 561 kJ of heat energy is taken in when solid carbon dioxide gas changes into its liquid form at the same temperature. Calculate the mass of carbon dioxide.

5 Liquid alcohol is provided with $5 \cdot 6 \times 10^5$ J of heat energy to change it to its gaseous form with no change in temperature. Calculate the mass.

6 0·83 MJ of energy is required to change 13 kg of solid gold to its liquid form with no change in temperature. Calculate the specific latent heat of fusion of gold.

7 A blacksmith heats a 42 kg iron bar, which changes state from solid to liquid with no change in temperature. The energy provided is $2 \cdot 66 \times 10^8$ J. Calculate the specific latent heat of fusion of iron.

8 A substance is heated. The graph below shows how the temperature of the substance changes over time.

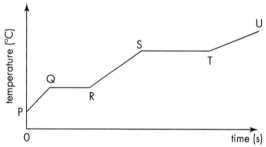

Describe what happens to the substance during section:

   a PQ   **b** QR   **c** RS   **d** ST   **e** TU

9 A block of ice of mass 2 kg is at a temperature of –5 °C. Calculate the total amount of heat energy required to convert the ice block to steam at 100 °C.

> **Hint** In order to calculate the total heat energy required, you must first calculate the energy required to change the **temperature** of the ice. Then calculate the energy required to change the **state** of the ice.

10 A 1·5 kW kettle is filled with 2·0 kg of water at 21 °C. The water is heated to its boiling point and steam is produced.

   a Calculate the heat energy required to bring the water in the kettle to its boiling point.

   b Upon reaching its boiling point, 0·2 kg of steam is produced. Calculate the energy required to make this change in state.

   c Calculate the total time taken to produce this amount of steam from water at 21 °C.

   d Explain why the actual time taken will be longer than the time calculated in **c**.

16 Gas laws and the kinetic model

Exercise 16A Pressure

1. Give the definition of pressure.

2. A box exerts a force of 80 N across an area of 1·5 m². Calculate the pressure.

3. Atmospheric pressure is 1×10^5 Pa. What force does the air exert on a roof that is 5 m wide and 7 m long?

4. A person is walking through heavy snow in a pair of snowshoes.

   The pressure on the snow is 4000 Pa when both feet are on the ground. The total force is 786 N. Calculate the area of one snowshoe.

5. A metal block with dimensions 0·7 m × 0·9 m × 1·3 m has a mass of 144 kg. Calculate:

   a the maximum pressure exerted when the block is lying flat on the ground

   b the minimum pressure exerted when the block is lying flat on the ground

6. A lorry has a mass of 3400 kg. There are six tyres on the vehicle. Each tyre has a contact area of 0·125 m². Calculate the pressure exerted on the road by the truck in kPa.

7. A 2700 kg elephant is standing in a forest. The area of one foot is 0·18 m². Calculate the total pressure exerted by the elephant on the forest floor in kPa.

8. A cube of side 3 cm is sitting flat on a bench. If the mass of the cube is 27 kg, what is the pressure on the bench?

1 Convert the following temperatures to Kelvin:

a 100 °C

b 200 °C

c −130 °C

d −22 °C

e −273 °C

2 Convert the following temperatures to degrees Celsius:

a 100 K

b 10 K

c 27 K

d 273 K

e 370 K

3 A sample of gas has a volume of 60 cm³ at a pressure of 1.5×10^5 Pa. Calculate the new volume of the gas when the pressure is increased to 4.8×10^5 Pa.

4 A syringe of volume 10 ml contains air at a pressure of 1.01×10^5 Pa. The volume of air in the syringe is reduced to 5 ml. Calculate the new pressure of the air inside the syringe.

5 At a certain height above the Earth, a helium weather balloon has a volume of 14 000 cm³ and a pressure of 9.80×10^4 Pa. As the balloon descends, the volume of the helium balloon decreases to 13 800 cm³. Assuming the temperature of the helium remains constant, calculate the new pressure.

6  A diver carries a cylinder of compressed air that has a volume of 40 litres and a pressure of 14 kPa. As the depth increases, the pressure of the compressed air increases to 15·6 kPa. Calculate the volume occupied by the compressed air at this pressure.

The pressure inside an unopened soft drinks can is 250 kPa at 20 °C. Assuming the volume of the can has not changed, calculate the pressure inside the can when it is cooled to 4 °C.

- Volume is constant so select the appropriate Gas Law:

$$\frac{P_1}{T_1} = \frac{P_2}{T_2}$$

- The temperature must be converted to Kelvin by adding 273:

$T_1 = 20 + 273 = 293$ K

$T_2 = 4 + 273 = 277$ K

- Now substitute the values into the equation:

$$\frac{250}{293} = \frac{P_2}{277}$$

$$P_2 = \frac{250}{293} \times 277$$

$$P_2 = 236 \text{ kPa}$$

7 Gas inside a sealed rigid container is initially at a pressure of $1 \cdot 68 \times 10^5$ Pa and a temperature of 31 °C. The temperature of the gas is raised to 55 °C. Calculate the new pressure of the gas in the container.

8 A cylinder of compressed gas has a pressure of $2 \cdot 8 \times 10^6$ Pa at a temperature of 21 °C. Calculate the pressure at a temperature of 4 °C.

9 The pressure of a car tyre is $2 \cdot 70 \times 10^5$ Pa at a temperature of 18 °C.

After a long journey, the pressure of the tyre is $2 \cdot 84 \times 10^5$ Pa. Assuming the volume of the tyre has not changed, calculate the temperature of the air inside the tyre.

10 A volume of 44 cm³ of air is at a temperature of 25 °C. Assuming the pressure remains constant, calculate the volume of air at a temperature of 47 °C.

11) The apparatus shown below is used to determine the relationship between the volume and temperature of a fixed mass of gas.

Hint Gas particles will only move faster if temperature is increased.

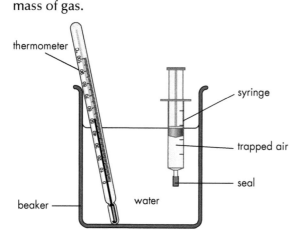

The water is heated, which in turn heats the trapped air inside the syringe. It is assumed that the trapped air is at the same temperature as the surrounding water. It is also assumed that the pressure of the trapped air remains constant throughout.

a The trapped air is at a volume of 19·4 cm³ when the temperature of the water is 22°C. Calculate the temperature of the trapped air at a volume of 23·9 cm³.

b Use the kinetic model to explain the change in volume as the temperature increases.

12) A volume of a fixed mass of gas is measured at five different temperatures. The pressure of the gas is kept constant throughout. The results are shown in the table below.

Volume (cm³)	0·200	0·207	0·214	0·221	0·228
Temperature (K)	280	290	300	310	320

a Using all the data, establish the relationship between the volume and the Kelvin temperature of a gas.

b Calculate the volume of the gas when the temperature is 62°C.

13) Describe an experiment to determine the relationship between the pressure and temperature of a fixed mass of gas at a constant volume.

Your description should include:

• a labelled diagram of the apparatus

• any measurements taken

14) A gas is initially at a pressure of 120 kPa, a volume of 22 litres and a temperature of 250 K. The pressure is then raised to 150 kPa and the temperature is increased to 400 K. Calculate the new volume of the gas.

15 A sealed syringe contains air at a pressure of 1.01×10^5 Pa, at a volume of 50 cm³ and a temperature of 20 °C. The air is heated to a temperature of 38 °C where the volume increases to 56 cm³. Calculate the new pressure of the air inside the syringe.

16 Inside a jet engine, air and fuel are drawn into the engine cylinder at a pressure of 1×10^5 Pa and a temperature of 18 °C.

The volume of the cylinder is 1000 cm³. The air is then compressed to a volume of 100 cm³, increasing the pressure to 14×10^6 Pa. Find the temperature of the air and fuel mixture after compression.

17 Wave parameters and behaviours

Exercise 17A Waves and energy

1 What is transferred by waves?

2 Consider the diagram below.

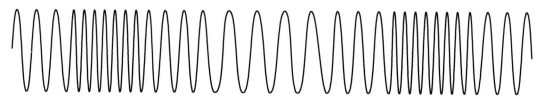

  a Is this wave longitudinal or transverse?

  b Explain your answer to **a**.

  c Give an example of this type of wave.

3 Consider the diagram below.

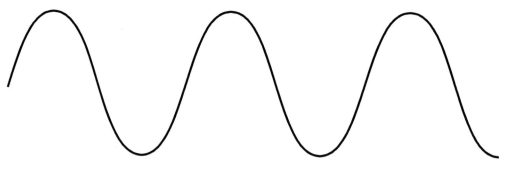

  a Is this wave longitudinal or transverse?

  b Explain your answer to **a**.

  c Give an example of this type of wave.

4 What is meant by the frequency of a wave?

5 Give a definition of the wavelength of a wave.

6 What is meant by the amplitude of a wave?

7 What is meant by the period of a wave?

8 What is meant by wave speed?

Exercise 17B Wave properties

1   Calculate the wavelength of the wave below.

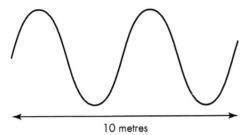

10 metres

2   Calculate the wavelength of the wave below.

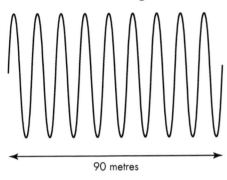

90 metres

3   Calculate the wavelength of the wave below.

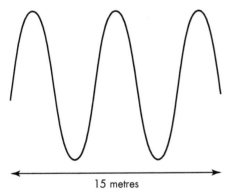

15 metres

4   What is the amplitude of this wave?

2 m

5 For the wave below:

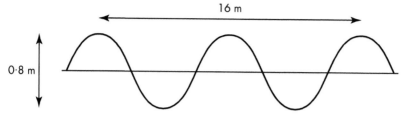

16 m

0·8 m

Calculate:

a the wavelength of the wave

b the amplitude of the wave

6 A girl stands on the end of a pier and counts the number of water waves in a given time. She counts 45 waves in 100 seconds. Calculate the frequency of the waves.

7 A wave machine in a swimming pool produces waves with a frequency of 0·88 Hz. How many waves are generated by the wave machine in 60 seconds?

8 Waves have a frequency of 4·2 Hz. An observer counts 28 waves passing a point. How long is the observer counting for?

9 A wave has a frequency of 47 Hz. Calculate the period.

> **Hint** The equation given in the relationships sheet is $T = \dfrac{1}{f}$.
>
> This means that if calculating frequency then $f = \dfrac{1}{T}$.

10 The period of a wave is 0·25 seconds. Calculate the frequency.

11 40 waves pass a point in 5 seconds. Calculate:

a the frequency of the waves

b the period of the waves

A pebble is thrown into a still pond, producing ripples at a rate of 4 waves per second.

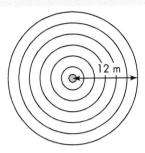

a   What is the frequency of the waves?

The definition of frequency is *the number of waves per second.*

So if there are 4 waves per second then frequency = 4 Hz

b   How many waves are represented in the diagram?

Each circle represents a wavefront or wave crest. So it is simply a matter of counting the number of wavefronts.

In this case, there are 6 wavefronts and therefore, 6 waves.

c   Calculate the wavelength of the waves.

If there are 6 waves in 12 metres then the length of one wave is:

$$\frac{12}{6} = 2 \text{ m}$$

12   A stone is thrown into a loch and produces ripples at a rate of 2 waves per second.

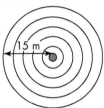

Calculate:

a   the frequency of the waves

b   the period of the waves

c   the wavelength of the waves

d   Another stone of a larger mass is thrown into the loch. What effect will this have on the amplitude of the waves?

Exercise 17C Speed, distance, time with waves

1 Water waves travel a distance of 20 m in 5 seconds. Calculate the wave speed.

2 A water wave is travelling at 2·5 ms⁻¹. How far will the wave travel in 40 seconds?

3 Sound waves are travelling through the air at 340 ms⁻¹. Calculate how long it takes the sound to travel a distance of 5000 m.

4 A dolphin uses a high frequency sound to communicate with other dolphins 15 km away. If the speed of sound in water is 1500 ms⁻¹, calculate the time taken.

5 A student hears thunder 6 seconds after she sees the lightning.

  a How far is the student from the lightning strike?

  b How long does it take light to travel this distance?

6 Microwave signals travel at the same speed as light in air. How long does the signal take to travel between two mobile phones 48 km apart?

7 The speed of sound in muscle is 1600 ms⁻¹. How far would sound travel in 0·26 ms?

8 An ultrasound pulse of frequency 3·5 MHz is transmitted into the womb of a pregnant mother. The pulse echo is detected 40 μs after it is transmitted. The speed of sound is 1500 ms⁻¹. Calculate the distance travelled by the ultrasound pulse.

9 A satellite is 36 000 km from both the transmitting station in Glasgow and the receiving station in Reykjavik. Calculate the time taken for radio waves to travel from Glasgow to Reykjavik.

10 A depth finder on a ship sends out a pulse of sound and detects the reflected pulse from the sea bed 0·43 seconds later. The speed of sound in water is 1500 ms⁻¹. Calculate the depth of the sea bed.

Exercise 17D Speed, frequency, and wavelength with waves

1 Water waves in a swimming pool have a wavelength of 0·5 m and a frequency of 3 Hz. Calculate the speed of the waves.

2 A tuning fork produces a note of frequency 440 Hz. The speed of the sound waves in air is 340 ms⁻¹. Calculate the wavelength.

3 A wave travelling at 12·5 ms⁻¹ has a wavelength of 4·9 mm. Calculate the frequency.

4 An X-ray machine emits waves with a wavelength of 5 nm. Calculate the frequency.

5 Calculate the wavelength of a radio wave of frequency 99 MHz.

6 A Wi-Fi signal in a person's home has a frequency of 3·6 GHz. The signals travel at 3×10^8 ms⁻¹. Calculate the wavelength.

7 A student is listening to music on their Bluetooth headphones.

The wavelength of the Bluetooth signal is 0·12 m. The signal travels at a speed of 3×10^8 ms⁻¹. Calculate the frequency of the Bluetooth signal.

8 A wave generator produces 50 waves per minute. The waves travel at a speed of 2·6 ms⁻¹.

Calculate:

a the wavelength

b the period of the waves

9 A wave travels a distance of 124 m in 6·7 seconds. The frequency of the wave is 83 Hz. Calculate the wavelength.

> **Hint** Sometimes there are multiple steps required in a wave's calculation. Try finding the value for the speed first.

10 Waves of frequency 7·1 GHz can cover a distance of 455 km in 52 ms. Calculate the wavelength.

1. Explain what is meant by diffraction of a wave.

2. A transmitter sends out radio waves and TV waves. A person living in a nearby valley can pick up the radio waves but not the TV waves. Explain why this is.

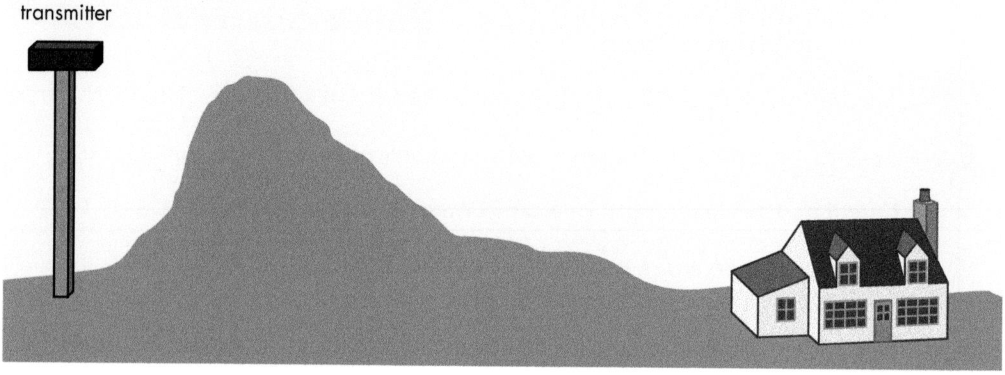

transmitter

Hint When waves diffract around obstacles, they continue to travel in the same direction without a change in wavelength.

3. Copy and complete the following diagrams:

   a

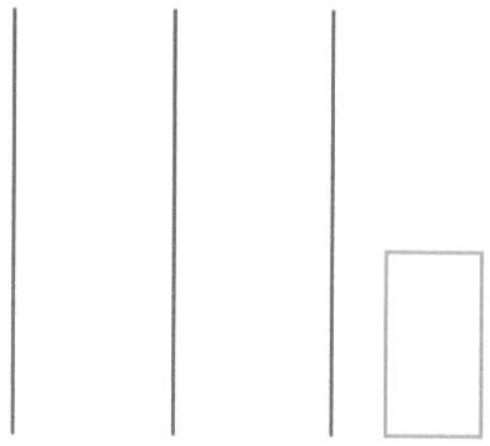

   b

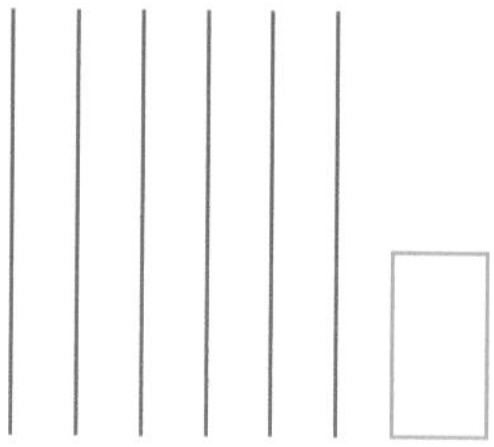

   c

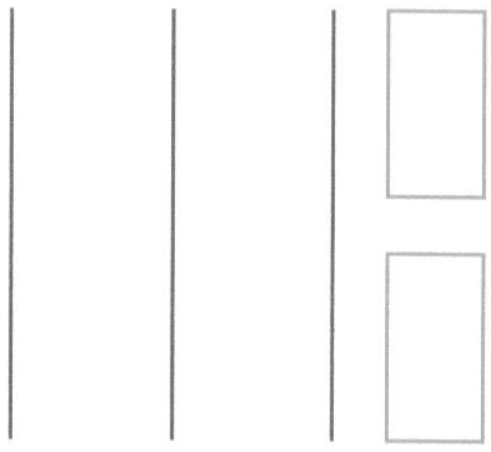

   d

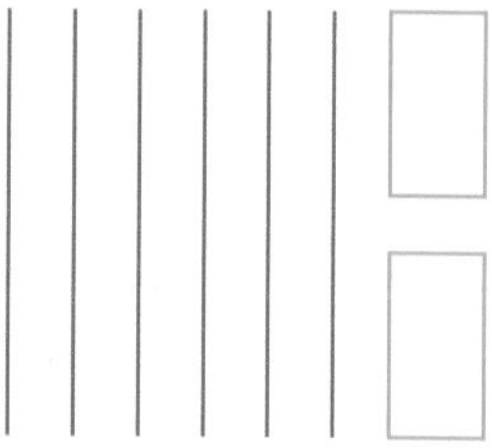

18 Electromagnetic spectrum

Exercise 18A The electromagnetic spectrum

1 What do all radiations in the electromagnetic spectrum have in common?

2 Which radiation in the electromagnetic spectrum has:

a the lowest frequency?

b the highest energy?

3 **a** Copy and complete the electromagnetic spectrum diagram shown below.

			Visible spectrum			Gamma rays

b Mark on your diagram the direction of increasing frequency and the direction of increasing wavelength.

4 Copy and complete the table below for each of the radiations in the electromagnetic spectrum.

Signal	Sources	Detectors	Applications
Radio waves	Stars, appliances	Aerial	Communications, radio and TV

5 Red light has a frequency of 4.29×10^{14} Hz. Blue light has a frequency of 7.5×10^{14} Hz. Show by calculation which colour of light has the longest wavelength.

19 Refraction of light

Exercise 19A Refraction

1 Explain what is meant by refraction of a wave.

2 Light travels from air to glass as shown below. Copy and complete the path of the ray of light through the glass block.

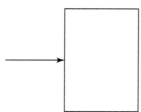

3 A ray of light passes from air to plastic as shown below.

> Hint   Always draw the normal line at 90° to the surface.

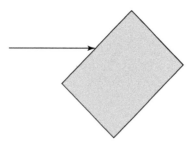

a Copy and complete the diagram to show the ray of light as it passes into the plastic block. Clearly label on your diagram the angle of incidence, the angle of refraction, and the normal.

b Describe the effect on speed and wavelength as the light enters the plastic.

4 A ray of light enters a glass prism as shown below.

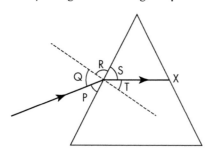

a Identify the angle on the diagram that represents:

  i the angle of incidence

  ii the angle of refraction

b Copy and complete the diagram to show the path of the ray of light as it leaves the prism at point X.

20 Nuclear radiation

Exercise 20A Ionising radiation

1 A radioactive source emits alpha particles. What is an alpha particle?

2 What is a beta particle?

3 State what is meant by a gamma ray.

4 What is meant by the term ionisation?

5 What is the effect of ionisation on neutral atoms?

6 State the range in air of alpha, beta and gamma radiation.

7 State the minimum amount of material that will absorb each of the three types of radiation.

8 Which of the three types of radiation, alpha, beta or gamma, produces the most ionisation?

9 A student is investigating the type of radiation being emitted from a radioactive source. The count rate with different materials placed in front of the detector is measured. The data gathered is shown below.

Material in front of detector	Count rate (counts per minute)
Air	1220
Sheet of paper	918
3 cm of aluminium	918
12 cm of lead	104

Which type(s) of radiation are emitted by this source?

10 A science technician is measuring the type of radiation emitted by three different sources. They measure the count rate with different materials placed in front of the detector. The results are shown in the table below.

Source	Count rate (counts per minute)			
	Air	Sheet of paper	3 cm aluminium	10 cm lead
A	550	278	19	19
B	876	876	876	14
C	230	12	13	12

a Which source emits only gamma radiation?

b What type of radiation does Source A emit?

Exercise 20B Activity

1 State what is meant by the activity of a radioactive source.

> **Hint** When giving a definition of a quantity, it is sometimes useful to look at the equation. In the case of $A = \dfrac{N}{t}$, what are the units for N and t?

2 What happens to the activity of a radioactive source as time passes?

3 A radioactive source produces 100 disintegrations per second. Calculate the activity.

4 A radioactive source produces 100 disintegrations in one minute. Calculate the activity in Bq.

5 Background radiation is measured at 30 counts per minute using a Geiger-Müller tube.

  a Convert this to becquerels.

  b List two sources of background radiation.

6 The activity of a radioactive source is 1 MBq. How many disintegrations would there be in one minute? Your answer should be in Bq.

7 Calculate the activity of a radioactive substance in which 252 000 atoms disintegrate in 2 minutes.

8 The activity of a sample of sea water is 14.5 kBq. Calculate the number of nuclear disintegrations that take place in the sample in 1 minute.

9 The activity of a radioactive source is 5 MBq. How many nuclear disintegrations occur in 6 hours?

10 The activity of the radioactive isotope carbon-14 is found to have reduced by 25 Bq. The total number of nuclear disintegrations is 4.5×10^{12} over a period of time. Calculate the time for this number of disintegrations to take place. Give your answer in years.

Exercise 20C Absorbed dose

> **Hint** Absorbed dose has the symbol D. It is easily confused with activity which has the symbol A.

1 Calculate the absorbed dose when a 2 kg mass absorbs 1 J.

2 Calculate the absorbed dose to a radiation worker, mass 80 kg, who absorbs 0.2 J of energy.

3 During radiotherapy over a period of several weeks, a 70 kg patient receives an absorbed dose of 10 Gy of radiation. How much energy did they receive?

4 When handling a radioactive source, a technician's hand receives an absorbed dose of 5 μGy. The mass of the hand is 365 g. Calculate the energy absorbed by the hand.

5 A small animal of mass 150 g absorbs 0.3 mJ of energy from a radioactive source. Calculate the absorbed dose.

6 A patient is receiving radiotherapy for a cancerous tumour. The tumour has a mass of 372 g and absorbs 6.6 mJ of energy. Calculate the absorbed dose.

7 A sample of tissue receives an absorbed dose of 240 mGy. The energy absorbed is 2.44×10^{-4} J. Calculate the mass of the tissue sample.

8 During a tooth X-ray, a patient's tooth receives an absorbed dose of 167 mGy and an energy of 4.86×10^{-5} J. Calculate the mass of the tooth.

9 Show by calculation that worker A, who has a mass of 60 kg and absorbs 120 mJ of energy from a radioactive source, receives a stronger dose than worker B, who has a mass of 70 kg and absorbs 126 mJ of energy.

10 A radiation worker receives an absorbed dose of 500 mGy of gamma radiation and 40 mGy of alpha radiation. Explain why it is not correct to say that the total absorbed dose is 540 mGy.

Exercise 20D Equivalent dose

1 A radiation worker receives an absorbed dose of 50 µGy of fast neutrons that have a radiation weighting factor of 10. Calculate the equivalent dose.

> **Hint** The radiation weighting factors can be found in the National 5 Physics data sheet.

2 A radiation worker receives an absorbed dose of 100 µGy of gamma radiation. Calculate the equivalent dose.

3 During an examination of cracks in metal, a worker is exposed to 25 µGy of gamma radiation. Calculate the equivalent dose.

4 X-rays produce an equivalent dose of 10 mSv for an absorbed dose of 10 mGy. Calculate the radiation weighting factor for X-rays.

5 Alpha particles produce an equivalent dose of 60 mSv and an absorbed dose of 3 mGy. Calculate the radiation weighting factor.

6 Alpha particles produce an equivalent dose of 29 µSv. If the radiation weighting factor is 20, calculate the absorbed dose.

7 A football player receives an X-ray with an equivalent dose of 6·5 mSv. Calculate the absorbed dose.

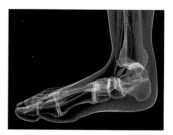

8 A radiation worker receives an absorbed dose of 120 µGy of fast neutrons and an absorbed dose of 60 µGy of beta particles.

a What is the equivalent dose for each radiation?

b Calculate the total equivalent dose.

9 During industrial testing for faults, a worker is exposed to an absorbed dose of 10 µGy of beta radiation and 25 µGy of gamma radiation. Calculate the total equivalent dose.

10 A radiation worker receives the following absorbed doses:

- 90 µGy of beta particles
- 130 µGy of gamma radiation
- 50 µGy of alpha particles

Calculate the total equivalent dose.

Exercise 20E Equivalent dose rate

1 State the exposure safety limits for:

  a the average annual background radiation in the UK

  b the annual effective dose limit for a member of the public

  c the annual effective dose limit for a radiation worker

2 A radiation worker receives an equivalent dose of 540 µSv during a 3-hour working week. Calculate the equivalent dose rate in µSv h⁻¹.

A person receives an equivalent dose of 5 mSv during a 15-minute CT scan. Calculate the equivalent dose rate in mSv per minute.

- $\dot{H}$ indicates the equivalent dose rate and H denotes the equivalent dose:

$$\dot{H} = \frac{H}{t}$$

- The final units are mSv per minute so there is no need to convert any of the units:

$$\dot{H} = \frac{5}{15}$$

  $\dot{H} = 0.3$ mSv per minute

3 A worker receives the following absorbed doses:

  - 250 µGy of gamma radiation
  - 130 µGy of fast neutrons
  - 110 µGy of slow neutrons

  a Find the total equivalent dose.

  b If the dose was received over a period of 8 hours, calculate the equivalent dose rate in µSv h⁻¹.

4 A radiation detector on an aircraft gives a reading of 0.1 nSv h⁻¹ during a 4-hour flight. Calculate the equivalent dose received by a passenger.

5 A person receives an equivalent dose of 20 µSv during a chest X-ray. If the equivalent dose rate is 5.4 µSv s⁻¹, calculate the time taken for the chest X-ray.

6 Following an earthquake in March 2011, a 15 m tsunami disabled the power supply and cooling system of three nuclear reactors in Fukushima, Japan. Following the disaster, a 20 km exclusion zone was set up to protect members of the public from overexposure to the radiation released. The equivalent dose in the exclusion zone over the first 2 weeks was 1 mSv. Calculate the equivalent dose rate in mSv y⁻¹.

7 During the Chernobyl disaster in 1986 the equivalent dose rate during explosion and meltdown was 5 Sv per minute. Calculate the equivalent dose during the first hour.

8 The equivalent dose rate limit at a nuclear power plant is 10 mSv y⁻¹. A worker receives, on average, an equivalent dose of 45 µSv per day. If the worker is on site for 195 days each year, show by calculation if the equivalent dose rate limit was exceeded.

Exercise 20F Half-life

1 What is meant by the half-life of a radioactive source?

2 Describe an experiment to measure the half-life of a radioactive source. Your description should include:

- the apparatus required
- the measurements taken
- how the half-life is calculated

3 A radioactive source has an activity of 160 Bq. The half-life of the source is 2 days. Assume the background count rate is negligible.

What is the activity after:

a 2 days?

b 4 days?

c 8 days?

4 A radioactive tracer is injected into the body with an activity of 320 Bq. The tracer has a half-life of 8 hours. What is the activity after 16 hours?

A radioactive source has a half-life of 6 hours. The initial activity of the source is 1000 Bq. Calculate the activity after 24 hours.

Initial activity = 1000 Bq

Half-life = 6 hours

This means that every 6 hours the activity will reduce by half.

Halve the activity each time and keep a track of how many half-lives there have been.

1000 Bq   500 Bq   250 Bq   125 Bq   62.5 Bq

So the final activity after 24 hours is 62·5 Bq.

5 A radioactive source has a half-life of 10 years. The source has an activity of 40 kBq in the year 2017. Calculate what the activity will be in the year 2067.

6 The activity of a source falls from 200 MBq to 12·5 MBq in a period of 8 years. Calculate the half-life of the source.

7 Calculate the half-life of a radioactive sample which has an initial activity of 330 kBq that drops to 82·5 kBq in 15 hours.

8 An alpha radiation source has an initial activity of 5200 counts per minute. After a period of 1 hour 15 minutes, the activity has fallen to 162·5 counts per minute. Calculate the half-life of the alpha radiation source.

9 The half-life of iodine-131 is 8 days. After a period of 24 days an iodine-131 source is found to have an activity of 7·5 counts per minute. Calculate the initial activity of the source.

10 A radioactive isotope has a half-life of 45 minutes. After a period of 3 hours the isotope has an activity of 22 MBq. Calculate the initial activity of the radioactive isotope.

11 A radioactive source has a half-life of 40 seconds. What fraction of the sample is left after:

a 40 seconds

b 80 seconds

c 320 seconds

12 A radioactive sample has a half-life of 8 minutes. How long has the sample been decaying for if the fraction left is:

a $\dfrac{1}{2}$ **b** $\dfrac{1}{8}$ **c** $\dfrac{1}{32}$

13 A radioactive source is to be used for medical treatment. The source has a half-life of 5 days.

The activity of the source when given to the patient must be 32·5 kBq. The source has an initial activity of 520 kBq at 9 am on September 5th. Calculate the time and date the source should be given to the patient.

14 The graph below shows the activity of a radioactive source over time.

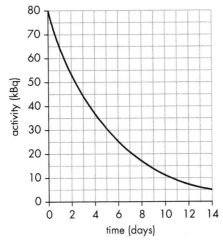

Use information from the graph to calculate the half-life of the source.

15 The graph below shows the activity of a radioactive source over time.

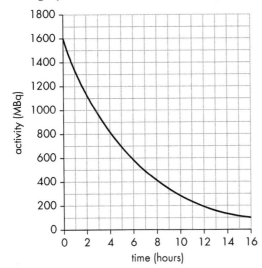

Use information from the graph to calculate the half-life of the source.

Exercise 20G Applications of nuclear radiation

1 Describe two applications of nuclear radiation.

2 State what is meant by the term nuclear fusion.

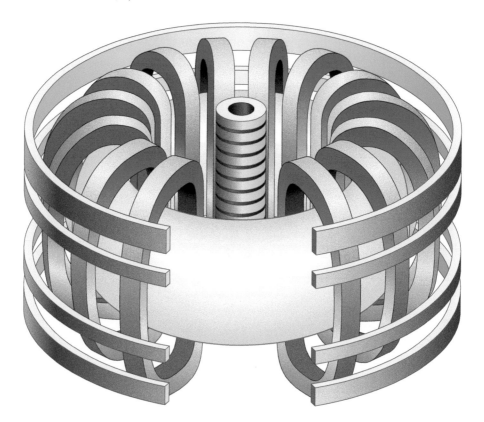

3 Describe what is meant by plasma containment.

4 State what is meant by the term nuclear fission.

5 Describe the process of a chain reaction in a nuclear fission reactor.

Notes

Notes